ANSYS® Tutorial

Release 2023R1

Structural & Thermal Analysis Using the ANSYS Mechanical APDL 2023R1 Environment

T0186165

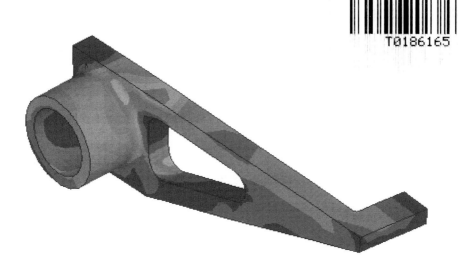

Kent L. Lawrence

Mechanical and Aerospace Engineering
University of Texas at Arlington

SDC
PUBLICATIONS

SDC Publications
P.O. Box 1334
Mission, KS 66222
(913) 262-2664
www.SDCpublications.com
Publisher: Stephen Schroff

Examination Copies:
Books received as examination copies are for review purposes only and may not be made available for student use. Resale of examination copies is prohibited.

Electronic Files:
Any electronic files associated with this book are licensed to the original user only. These files may not be transferred to any other party.

Trademarks:
ANSYS, ANSYS Mechanical, ANSYS Multiphysics, Workbench, and any and all ANSYS, Inc. product and service names are registered trademarks or trademarks of ANSYS, Inc. or its subsidiaries located in the United States or other countries. All other trademarks or registered trademarks are the property of their respective owners.

The author and publisher of this book have used their best efforts in preparing this book. The efforts include the testing of the tutorials to determine their effectiveness. However, the author and publisher make no warranty of any kind, expressed or implied, with regard to the material contained in this book. The author and publisher shall not be liable in any event for incidental or consequential damages in connection with the use of the material contained herein.

ISBN-13: 978-1-63057-613-4
ISBN-10: 1-63057-613-1

Printed and bound in the United States of America.

TABLE OF CONTENTS

LESSON 3 – AXISYMMETRIC PROBLEMS

LESSON 4 – THREE-DIMENSIONAL PROBLEMS

LESSON 5 – BEAMS

LESSON 6 – SHELLS

LESSON 7 – HEAT TRANSFER, THERMAL STRESS

LESSON 8 – SELECTED TOPICS

REFERENCES

PREFACE

The eight lessons in the **ANSYS Tutorial** introduce the reader to effective engineering problem solving through the use of this powerful finite element analysis tool. Topics include trusses, plane stress, plane strain, axisymmetric problems, 3-D problems, beams, plates, conduction/convection heat transfer, and thermal stress. It is designed for practicing and student engineers alike and is suitable for use with an organized course of instruction or for self-study.

The ANSYS software is one of the most mature, widely distributed and popular commercial FEM (Finite Element Method) programs available. In continuous use and refinement since 1970, its long history of development has resulted in a code with a vast range of capabilities. This tutorial introduces a basic set of those to the reader. An extensive on-line help system, a number of excellent tutorial web sites, and other printed resources are available for further study.

I am most appreciative of the long-time support of research and teaching efforts in our universities provided by ANSYS, Inc. Heartfelt thanks also go to UTA graduate students past and present who read portions of the material as well as test-solved a number of the problems. I also appreciate the comments and suggestions provided by Prof. Wen Chan (deceased) and by Dr. C. Y. Lin.

Stephen Schroff of SDC Publications has been very helpful to me in my efforts during the preparation of this material. To Stephen I am most indebted as well as to SDC staff for careful assistance with the manuscript.

Special thanks go as usual to Carol Lawrence, ever supportive and always willing to proof even the most arcane and cryptic stuff.

Please feel free to point out any problems that you may notice with the tutorials in this book. Your comments are welcome at **lawrence@uta.edu** and can only help to improve what is presented here.

Kent L. Lawrence

NOTES:

Introduction

I-1 OVERVIEW

Engineers routinely use the **Finite Element Method (FEM)** to solve everyday problems of **stress, deformation, heat transfer, fluid flow, electromagnetics**, etc. using commercial as well as special purpose computer codes. This book presents a collection of tutorial lessons for **ANSYS Mechanical APDL**, one of the most versatile and widely used of the commercial finite element programs.

The lessons discuss linear static response for problems involving truss, plane stress, plane strain, axisymmetric, solid, beam, and plate structural elements. Example problems in heat transfer, thermal stress, mesh creation and transferring models from CAD solid modelers to ANSYS are also included.

The tutorials progress from simple to complex. Each lesson can be mastered in a short period of time, and Lessons 1 through 7 should all be completed to obtain a thorough understanding of basic ANSYS structural analysis. ANSYS APDL running on Windows was used in the preparation of this text.

I-2 THE FEM PROCESS

The finite element process is generally divided into three distinct phases:

1. **PREPROCESSING** – Build the FEM model.

2. **SOLVING** – Solve the equations.

3. **POSTPROCESSING** – Display and evaluate the results.

The execution of the three steps may be performed using a **batch** process, an **interactive** session, or a **combination** of batch and interactive processes. These tutorials cover all three approaches.

I-3 THE LESSONS

A short description of each of the ANSYS Tutorial lessons follows.

Lesson 1 utilizes simple, two-dimensional **truss** models to introduce basic ANSYS concepts and operation principles.

Lesson 2 covers problems in **plane stress** and **plane strain** including a discussion of determining stress concentration effects. Geometry from a CAD solid modeler is imported into ANSYS for FEM analysis.

Lesson 3 discusses the solution of **axisymmetric** problems wherein geometry, material properties, loadings and boundary conditions are rotationally symmetric with respect to a central axis.

Lesson 4 introduces three-dimensional **tetrahedron** and **brick** element modeling and the various options available for the analysis of complex geometries. Importing models from CAD systems is also discussed.

Lesson 5 treats problems in structural analysis where the most suitable modeling choice is a **beam** element. Both two-dimensional and three-dimensional examples are presented.

Lesson 6 covers the use of **shell** (plate) elements for modeling of structural problems of various types. Plates of Orthotropic and Composite Materials are covered.

Lesson 7 presents tutorials in the solution of conduction/convection problems of **heat transfer** in solids. It concludes with an analysis of a case of **thermal stress** in an object whose temperature distribution has been determined by an ANSYS thermal analysis.

Lesson 8 covers additional short topics of interest to the ANSYS user. These include problems with multiple load cases, determining natural frequencies of structural systems, and others.

The lessons are necessarily of varying length and can be worked through from start to finish. Those with previous ANSYS experience may want to skip around a bit as needs and interests dictate.

I-4 THE *ANSYS* INTERFACE

To start ANSYS on a Windows-based system, select **Mechanical APDL (ANSYS)** or **Mechanical APDL Product Launcher** from the desktop menu shown below.

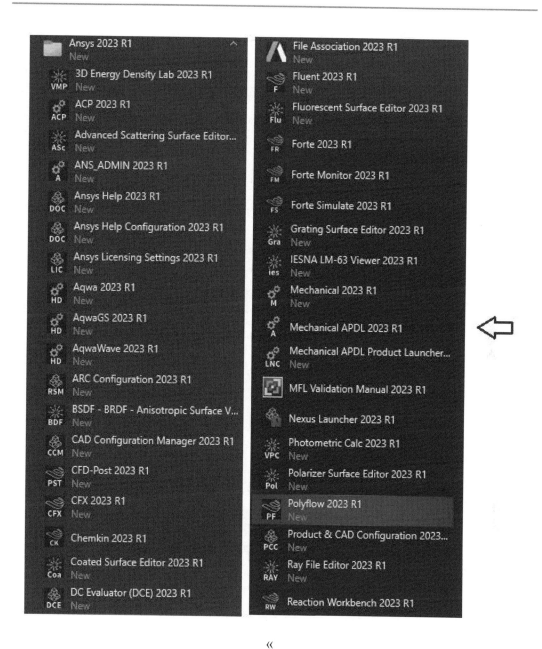

Figure I-1 ANSYS desktop menu.

The **Product Launcher** provides an opportunity to select a number of system options including the default working directory. This is very helpful in some network startup situations.

Figure I-2 shows the **ANSYS interface** with **Utility Menu** across the top followed by the ANSYS **Command Input** area, **Main Menu**, **Graphics** display, and ANSYS **Toolbar**.

Figure I-2 ANSYS Start Screen.

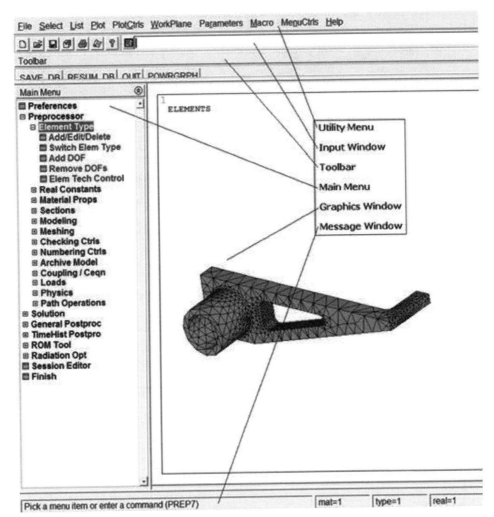

Figure I-3 ANSYS Interface.

Be sure to check the **Message Window** for prompts if you need to know what is the next action you should take to complete an operation.

Program actions can be initiated in a number of ways: by **text file input**, by **command inputs**, or by **menu picks**. In some cases (creating lists of quantities, for example) there may be more than one menu pick sequence that produces the same results. These ideas are explored in the lessons that follow.

I-5 ANSYS HELP

A wealth of on-line help information is provided with the ANSYS software. Select **ANSYS Help > Help Topics**, and the following information screen is displayed.

Figure I-4 ANSYS help facility.

The help contents shown include a number of analysis guides including **structural analysis** and **thermal analysis**

Select **Mechanical APDL > Structural**

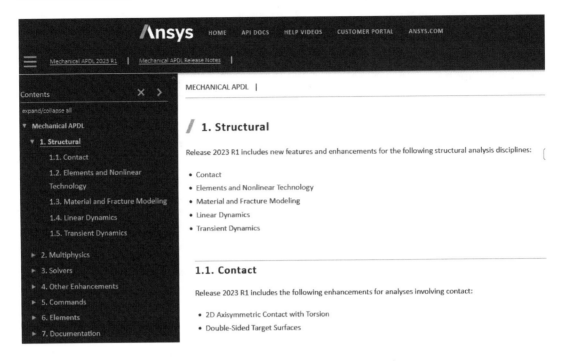

Figure I-5 Structural Analysis Help Topics.

Much can be learned from the theory, element descriptions and solution of the problems described in the **Verification Manual**. Each VM problem provides a copy of the problem text file and it can be used to duplicate the results discussed. Select **Static Structural Analysis > 2.5 Where to Find Other Examples**

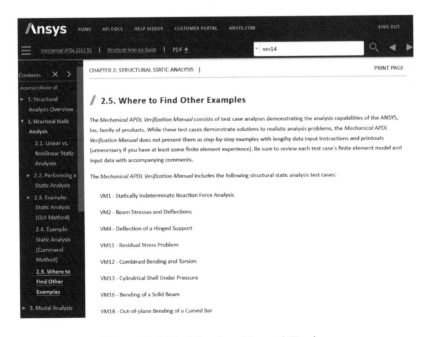

Figure I-6 Verification Manual Topics.

I-6 SUMMARY

The tutorials in this book explain some of the many engineering problem solution options available in ANSYS. It is important to remember that while practically anyone can learn which buttons to push to make the program work, the user must first possess an adequate engineering background, then give careful consideration to the assumptions inherent in the modeling process and finally give an equally careful evaluation of the computed results.

In short, model building and output evaluation based upon sound fundamental principles of engineering and physics is the key to using any program such as ANSYS **correctly**.

NOTES:

Lesson 1

Trusses

1-1 OVERVIEW

Simple truss structural analysis is used in this lesson to introduce the FEA framework and the general concepts that will be used in the lessons that follow. Truss modeling may not be your primary goal in undertaking these lessons, but it provides us a convenient vehicle for beginning a study of finite element methods with ANSYS. The three tutorials explore:

- ♦ Creating a text file FEM model definition.

- ♦ Problem solving interactively using the ANSYS interface.

- ♦ Combining text file input and interactive solution methods.

In addition, a number of additional truss modeling possibilities and formulations are demonstrated.

1-2 INTRODUCTION

Some structures are built from elements connected by pins at their joints. These **two force elements** can carry only axial tensile or compressive forces and stresses.

Figure 1-1 Two-force element.

In addition, other structures built of long slender members that are welded or bolted together may carry such small bending and torsion loads that they can be accurately analyzed by using a pin-jointed model.

ANSYS provides a comprehensive library of elements for use in the development of models of physical systems. While engineering problems always arise from consideration of some real-life 3D situation, 2D models are oftentimes sufficient for describing the behavior and therefore are widely used. The ANSYS two-force element **LINK180** is used for truss models.

1-3 SHELF TRUSS

Figure 1-2 shows a steel shelf we wish to analyze to determine the **maximum stress** and **deflection** of the support structure. To do this, we consider a two-dimensional model of the pin-jointed structure that supports each end of the shelf.

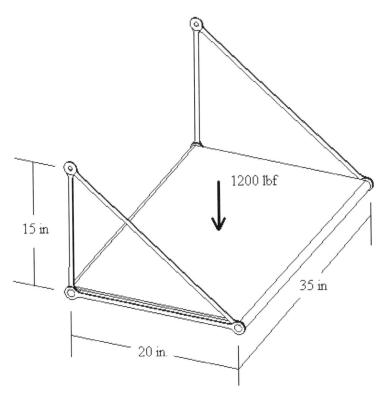

1200 lbf

15 in

35 in

20 in.

Figure 1-2 Shelf.

Depending on the connection details, there could be some twisting of the horizontal truss element of our problem. Twisting is ignored here but can be considered in a more detailed model later if necessary. Good engineering practice often involves first analyzing the simplest model that can address the major questions being considered and subsequently developing more complex models as understanding of the problem grows. Figure 1-3 displays the FEM model of the shelf support.

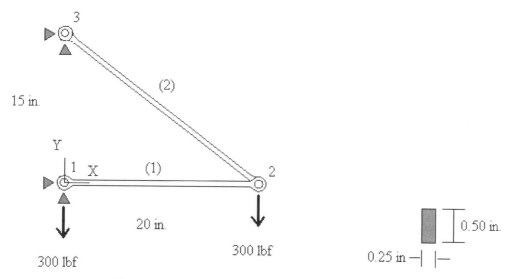

Figure 1-3 Simple truss and element cross section.

The XYZ coordinate system is the **global axis system** for the problem. The location and orientation of this reference axis system is selected by the engineer in setting up the model. **Nodes 1, 2,** and **3** are points where **element (1)** and **element (2)** join each other and/or the support structure. Elements (1) and (2) carry only axial loads, and ANSYS link1 type elements are used to model their behavior.

The small triangles are used to indicate a **displacement constraint**, no motion in this case. We assume a uniform distribution of the 1200 lbf shelf load to its four supports. The 300 lbf force at node 1 is reacted directly by the support at that node but is shown anyway for completeness.

The **area** of the structural elements is 0.25 x 0.50 = **0.125 in²**, and the values of the material properties we will use for **steel** are: **elastic modulus, E = 30 x 10⁶ psi; Poisson's ratio,** $\nu = 0.27$.

A free body at node 2 shows that element (1) will be in **compression** and element (2) will be in **tension**. We can use this observation to check the computed results.

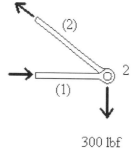

Figure 1-4 Free body of node 2.

We can now list the seven input quantities that are needed to characterize a typical finite element model. These items are defined during the **Preprocessing** phase.

Items that Constitute a Typical Finite Element Model

1. The types of elements used in assembling the model.

2. Element geometric properties such as areas, thicknesses, etc.

3. The element material property values.

4. The location of the nodes with respect to the user defined XYZ system.

5. A list showing which elements connect which nodes.

6. A definition of the nodes with displacement boundary conditions.

7. The loadings, their magnitudes, locations and directions.

In the tutorial below we define the model for this problem by creating a **text file** that contains the seven groups listed above. The **Preprocessing** step is then performed when ANSYS reads the text file. The **Solution** and **Post Processing** steps are performed interactively.

1-4 TUTORIAL 1A – SHELF TRUSS

Follow the steps below to analyze the truss model. The tutorial is divided into separate Preprocessing, Solution, and Postprocessing steps.

Objective: Determine the maximum stress and deflection. Check for buckling.

PREPROCESSING

1. Create Input Data File - Use a text editor such as Notepad to prepare a file that contains the information shown below. Lines beginning with '/' contain **ANSYS operations** to be performed. Other lines contain **input data** or comments. A **comment** begins with '!'.

Save the file as **T1A.txt** or another convenient name. To obtain meaningful and easily interpreted results, we **must use a consistent set of units** for all input quantities, **inches** and **pounds force** in this case.

You can **add your name** to the title 'Simple Truss' to identify the plotted output.

```
/FILNAM,truss
/title, Simple Truss

/prep7

et, 1, link180              ! Element type; no.1 is link180

sectype, 1, link           ! Type of cross section is link

secdata, 0.5               ! Cross sectional area = 0.5 sq in

mp, ex, 1, 3.e7            ! Material Properties, E for material no. 1
mp, prxy, 1, 0.27         ! & poisson's ratio for material no.1

n, 1,  0.0,  0.0, 0.0     ! Node 1 is located at (0.0, 0.0, 0.0)
n, 2, 20.0,  0.0, 0.0
n, 3,  0.0, 15.0, 0.0

en, 1, 1, 2               ! Element Number 1 connects nodes 1 & 2
en, 2, 2, 3

d, 1, ux, 0.              ! Displacement at node 1 in x-dir is zero
d, 1, uy, 0.              ! Displacement at node 1 in y-dir is zero
d, 3, ux, 0.
d, 3, uy, 0.

f, 1, fy, -300.           ! Force at node 1 in y-direction is -300.
f, 2, fy, -300.           ! Force at node 2 in y-direction is -300.
```

2. Select **Mechanical APDL 2023** from the ANSYS 2023 menu.

Figure 1-5 Desktop menu.

The **ANSYS interface** shown in the figure below includes the **Utility Menu** across the top followed by the ANSYS **Command Input** area, the **Main Menu**, the **Graphics** display, and the ANSYS **toolbar**.

The **Jobname** and **Title** to be used for this analysis are contained in the first two lines of the text file. ANSYS will use the file name for any files created or saved during this computation.

To change from **black** background to **white** use: **PlotCtrls > Style > Colors > Reverse Video**

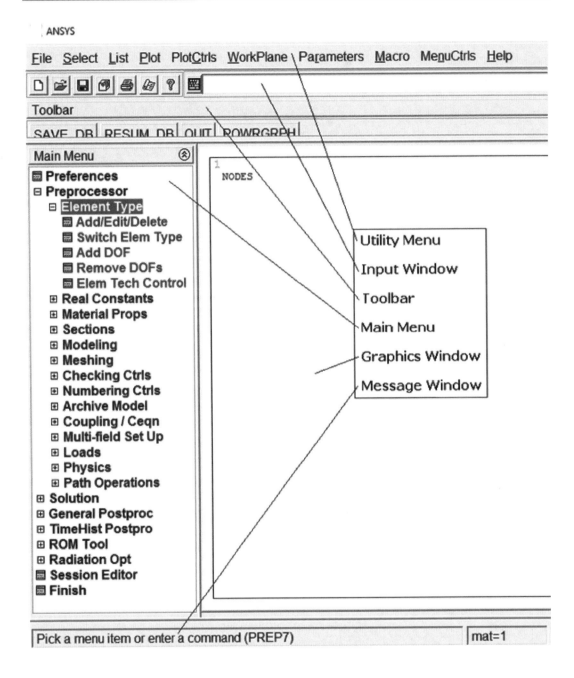

Figure 1-6 ANSYS interface components.

In the tutorials in this book we will make extensive use of the Utility Menu, Main Menu and Graphics Window. It is easy to overlook the **message window** at the bottom left corner of the screen. When in doubt as to what to do next, check here first.

Set the **working directory** you want to use for this analysis.

3. Utility Menu > File > Change Directory (search files tree) **> OK**

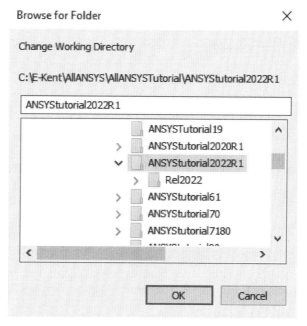

Figure 1-7 Set the working directory.

(If you want to use a different Jobname and/or a different title, use **Utility Menu > File > Change Jobname …** and **Utility Menu > File > Change Title …**)

Next we will read the text data file you prepared for this problem into ANSYS.

4. Utility Menu > File > Read Input From   (Find and select the file **T1A.txt** that you prepared earlier.)

After the file is read, we will **display the model graphically** and check to see that the model is correctly defined.

Figure 1-8 File menu.

Graphical image manipulation can be performed using the Pan, Zoom, Rotate pull-down menu or the equivalent icons on the right border of the graphics screen as well as **mouse wheel zoom. (Utility Menu > PlotCtrls > Pan, Zoom, Rotate …)**

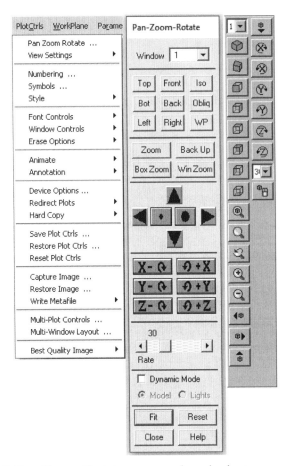

Figure 1-9 Pan, Zoom, Rotate menu and equivalent on-screen icons.

Turn on node and element numbering graphics options.

5. Utility Menu > PlotCtrls > Numbering > Node numbers ON, Element / Attrib numbers > Element numbers > OK

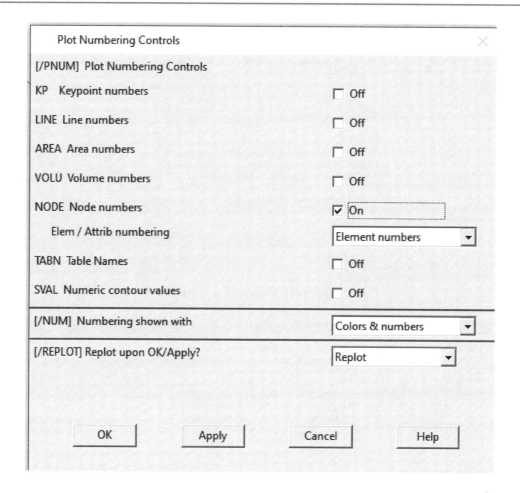

Figure 1-10 Numbering controls.

Set up the graphics to display boundary conditions and loads.

6. Utility Menu > PlotCtrls > Symbols > All Applied B.C.'s > OK

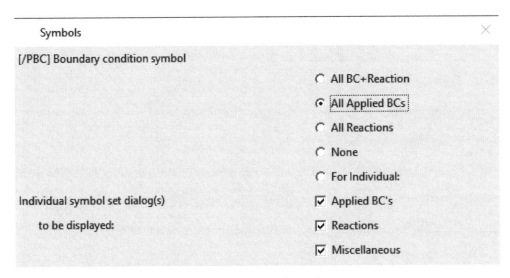

Figure 1-11 Symbols options.

(You may have to reset or adjust these plotting options from time to time during the analysis to obtain the graphic information you want to see.)

7. Fit 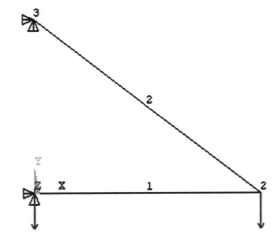 **> Plot > Elements** (use Fit or Zoom, Pan controls as needed including **middle mouse wheel** for dynamic zoom.)

The plot of your model is now displayed as shown in the figure to the right.

Figure 1-12 Plot of model.

SOLUTION

8. Main Menu > Solution > Solve > Current LS > OK   (Current Load Step)

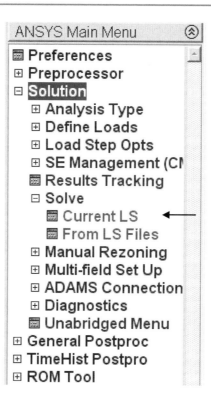

Figure 1-13 Solution menu.

The **/STATUS Command** window displays the problem parameters and the **Solve Current Load Step** window is shown. Check the solution options in the /STATUS window and if all is OK, in the **Solve Current Load Step** window, select **OK** to compute the solution. (Multiple load steps may be used in the solution of nonlinear problems.)

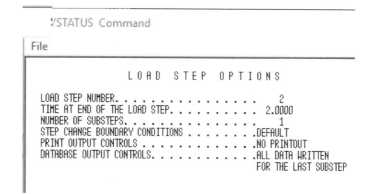

Figure 1-14 Solution options.

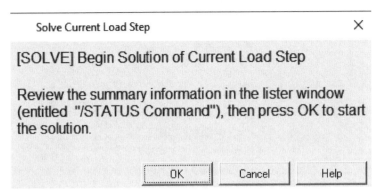

Figure 1-15 Solve.

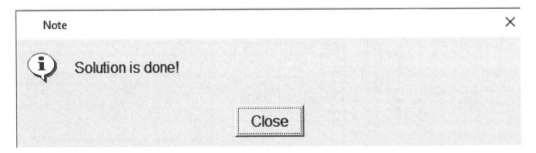

Figure 1-16 Solution completion.

Once the solution is complete, **Close** the **Note** and **STATUS** windows. The **Raise**
Hidden button near the top of the screen can be used to bring the STATUS window
(or any other hidden window) to the foreground.

POSTPROCESSING

We can now **plot the results** of this analysis and also **list the computed values**.

**9. Main Menu > General Postproc > Plot Results > Deformed Shape > Def. + Undef.
> OK**

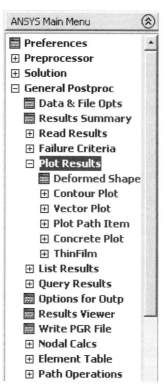

Figure 1-17 Postprocessing options.

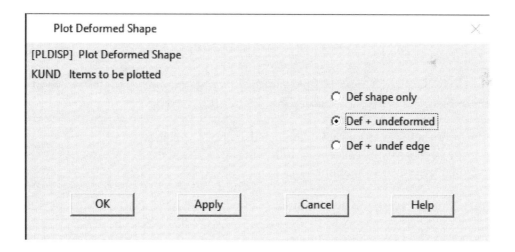

Figure 1-18 Plot deformed shape options.

The following graphic is created.

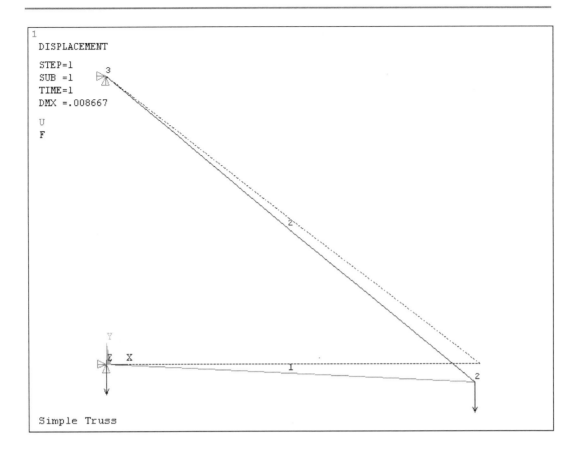

Figure 1-19 Deformed shape plot.

Use **PlotCtrls** > **Hard Copy** > **To Printer** to print graphics. (**PlotCtrls** > **Style** > **Background,** uncheck **Display Picture Background** if need be.)

It is often very useful to **animate** the deflected shape to visualize how the structure is behaving. Make the following selections from the utility menu to create an animation.

10. Utility Menu > PlotCtrls > Animate > Deformed Shape > OK

Use the Raise Hidden button and Animation Controller to stop animation. A list of the computed results can be obtained from the **General Postprocessor Menu**.

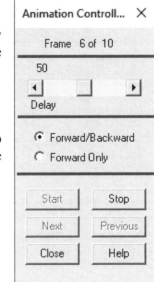

Figure 1-20 Animation Controller.

Next we will **Read** and **List calculated numerical** results.

Read results.

11. Main Menu > General Postproc > Read Results > First Set

12. Main Menu > General Postproc > List Results > Nodal Solution > DOF Solution > Displacement vector sum > OK

□ **General Postproc**
  ▦ **Data & File Opts**
  ▦ **Results Summary**
  □ **Read Results**
    ▦ **First Set**
    ▦ **Next Set**
    ▦ **Previous Set**
    ▦ **Last Set**
    ▦ **By Pick**
    ▦ **By Load Step**
    ▦ **By Time/Freq**
    ▦ **By Set Number**
  ⊞ **Failure Criteria**
  ⊞ **Plot Results**
  ⊞ **List Results**

```
PRINT U    NODAL SOLUTION PER NODE

 ***** POST1 NODAL DEGREE OF FREEDOM LISTING *****

 LOAD STEP=     1  SUBSTEP=      1
  TIME=    1.0000       LOAD CASE=   0

 THE FOLLOWING DEGREE OF FREEDOM RESULTS ARE IN
GLOBAL COORDINATES

    NODE     UX          UY          UZ          USUM
       1   0.0000      0.0000      0.0000      0.0000
       2 -0.21333E-02-0.84000E-02 0.0000      0.86667E-02
       3   0.0000      0.0000      0.0000      0.0000

MAXIMUM ABSOLUTE VALUES
NODE          2           2           0           2
VALUE  -0.21333E-02-0.84000E-02 0.0000      0.86667E-02
```

Node 2 is the only node allowed to move in this simple model, and from the above we see that the **maximum deflections** are 0.0021 inches **left** and 0.0084 inches **down**. Although no design specifications were mentioned in the problem definition, the shelf support structure seems to be pretty stiff.

Let's check the stresses.

13. Main Menu > General Postproc > List Results > Element Solution > Line Element Results > Element Results > OK

```
PRINT ELEM ELEMENT SOLUTION PER ELEMENT

 ***** POST1 ELEMENT SOLUTION LISTING *****

 LOAD STEP    1  SUBSTEP=     1
  TIME=    1.0000        LOAD CASE=  0

EL=      1 NODES=       1       2 MAT=      1  XC,YC,ZC=  10.00
0.000     0.000     AREA= 0.12500              LINK180
  FORCE= -400.00      STRESS= -3200.0     EPEL=-0.10667E-03
  TEMP=   0.00   0.00  EPTH=  0.0000

EL=      2 NODES=       2       3 MAT=      1  XC,YC,ZC=  10.00
7.500     0.000     AREA= 0.12500              LINK180
  FORCE= 500.00       STRESS=  4000.0     EPEL= 0.13333E-03
  TEMP=   0.00   0.00  EPTH=  0.0000
```

This file gives the input parameters for each element as well as the computed results. **FORCE** is the **axial force**, and **STRESS** is the **axial stress**. Element (1) is in compression (-400 lbf, -3200 psi) and element (2) is in tension (500 lbf, 4000 psi) as anticipated.

The stress values are well below yield strengths of commonly used steels, but since element (1) is in **compression**, there is the possibility that it could fail by **buckling**. This possibility should be checked by a hand calculation.

In the X-Y plane the horizontal link can be considered a pinned end column. Its flexural inertia for buckling in the X-Y plane is $I = 0.0026$ in^4. The Euler pinned-end buckling load is $P_{cr} = \pi^2 \, EI/L^2 = 1924$ lbf which is greater than the 400 lbf load it is carrying, so it seems OK.

However, the horizontal link could also buckle in the X-Z plane, that is, in a direction normal to the plane of analysis. The end conditions are not pinned for this mode of deformation but probably not quite completely fixed either, so this would require further investigation on the part of the stress analyst.

To obtain the **support reactions**

14. Main Menu > General Postproc > List Results > Reaction Solution > All items > OK

At any point you can **save your work** using the ANSYS **binary file format**. To do this

15. Utility Menu > File > Save as Jobname.db   (Or >Save as *supply a new name***)**

Your work is saved in the working directory using the current Jobname. The problem is stored using the default ANSYS file format, and the text file you used to define the problem is really no longer needed. **Reload** this model using **File > Resume Jobname**.

To begin a **NEW PROBLEM** use:

16. Utility Menu > File > Clear & Start New ... > OK > Yes
Otherwise data from the old problem may contaminate the new one.

1-5 TUTORIAL 1B – MODIFIED TRUSS

Suppose we wish to evaluate the performance of a composite tube as a replacement for the steel tension member, element (2) in the truss model above. The elastic modulus of the composite tube material is **1.2 x 10^7 psi**, its Poisson's ratio is **0.3** and the tube has a cross sectional area of **0.35 in^2**.

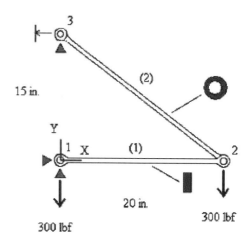

Figure 1-21 Modified truss.

The model must now include two different materials and two different cross sectional geometric properties. These are included in the text file description of the model using the (archived but functional) **mat** and **real** commands to specify which property to use when creating each element.

We also want to know how the structure is affected by a **0.01 inch movement of the upper support point to the left**. This movement can be incorporated in the model as a nonzero displacement boundary condition.

Finally, in doing this computation we will include the Solution and Postprocessing instruction in the text file and not perform those steps interactively as we did in the previous tutorial. The appropriate text file is shown below.

1. Prepare Input File

```
/FILNAM,T1B
/title, Simple Truss w Composite Tube
/prep7

et, 1, link180           ! Element type; no.1 is link180

! Material 1
mp, ex, 1, 3.e7          ! Material Properties: E for material no. 1
mp, prxy, 1, 0.27        ! & poisson's ratio for material no.1

! Material 2
mp, ex, 2, 1.2e7         ! Material Properties: E for material no. 2
mp, prxy, 2, 0.3         ! & poisson's ratio for material no.2


! Cross sectional areas (Also see 'real const' note at end of this chapter)
! First area
r, 1, 0.125             ! Real Constant number 1 is 0.125

! Second area
r, 2, 0.35              ! Real Constant number 2 is 0.35

n, 1,  0.0,  0.0, 0.0   ! Node 1 is located at (0.0, 0.0, 0.0)
n, 2, 20.0,  0.0, 0.0
n, 3,  0.0, 15.0, 0.0

! Set properties before creating elements
mat, 1
real, 1
en, 1, 1, 2             ! Element Number 1 connects nodes 1 & 2
```

```
mat, 2
real, 2
en, 2, 2, 3                 ! Element Number 2 connects nodes 2 & 3

d, 1, all, 0.              ! All Displacements at node 1 are zero
d, 3, ux, -0.01            ! Horizontal Displacement at node 3 is -0.01
d, 3, uy,  0.              ! Vertical Displacement at node 3 is zero


f, 1, fy, -300.           ! Force at node 1 in y-direction is -300.
f, 2, fy, -300.           ! Force at node 2 in y-direction is -300.

finish

/solu                     ! Select static load solution
antype, static
solve
save
finish
```

2. Utility Menu > File > Read Input from

3. Postprocessing Follow the steps given in Tutorial 1A to find the new model results. The displacements of node are found to be ux = -0.21333E-002 and uy = 0.55286E-002 inches.

1-6 TUTORIAL 1C – INTERACTIVE PREPROCESSING

Preprocessing by a text file defining the problem was shown in Tutorials 1A and 1B. The **Preprocessing** can also be performed in an **interactive session**. Next we redo Tutorial 1B, but perform the model creation step using interactive menu picks and data entry.

1. Start ANSYS and set the **jobname** to **Tutorial1C** or something else easy to remember, set the desired **working directory**, and set **Title** to something descriptive.

2. Main Menu > Preferences > Structural > OK

(Here we are setting the type of problem to be solved and method used. Actually the defaults are Structural and h-Method, so we do this only to introduce the ideas. More on this later.)

Indicate that we will use **the link180** element for modeling.

3. Preprocessor > Element Type > Add/Edit/Delete > Add > Structural   3D finit stn 180 > OK > Close

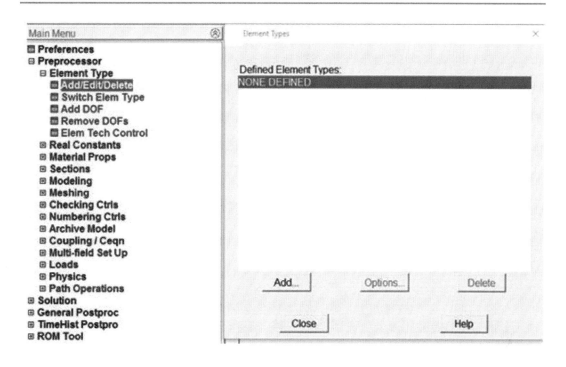

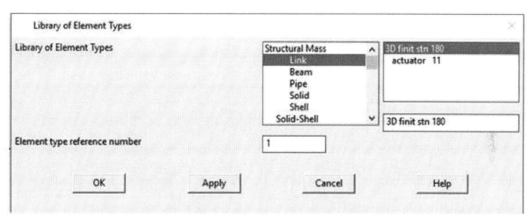

Figure 1-22 Select element type.

A problem may have more than one kind of element. This establishes Element type reference number 1 as the 3D finit stn, link180. (If we had some beams, we would add reference number 2 as a beam, etc. Each different type needs an identifying number.) Enter the element cross sectional areas for the rectangular bar and the circular tube.

4. Main Menu > Preprocessor > Sections > Link > Add > OK

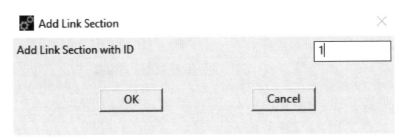

Figure 1-23 Section constant for link 1.

Element Type Reference No. 1

Create Section Constant Set No. 1

Enter **Link Area** = **0.5** for Section constant set 1 **OK > Close**

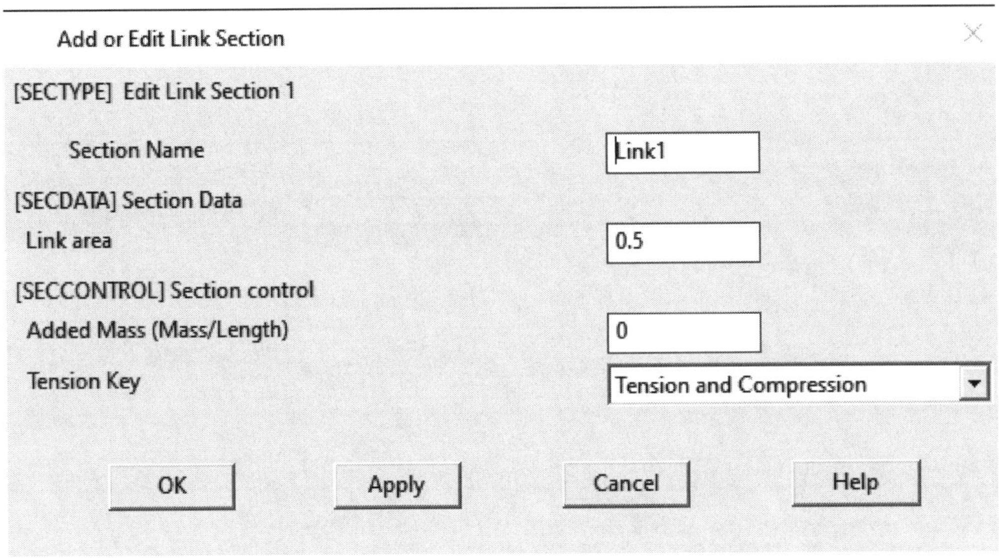

Figure 1-24 Section constant set No. 1.

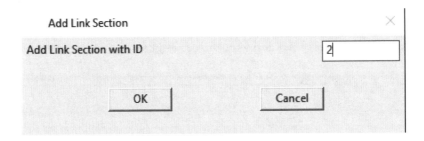

Figure 1-25 Section constant for link 2.

Add the second element size.

Element Type Reference No. 1

Add link section No. 2

Enter **AREA** = **0.35**

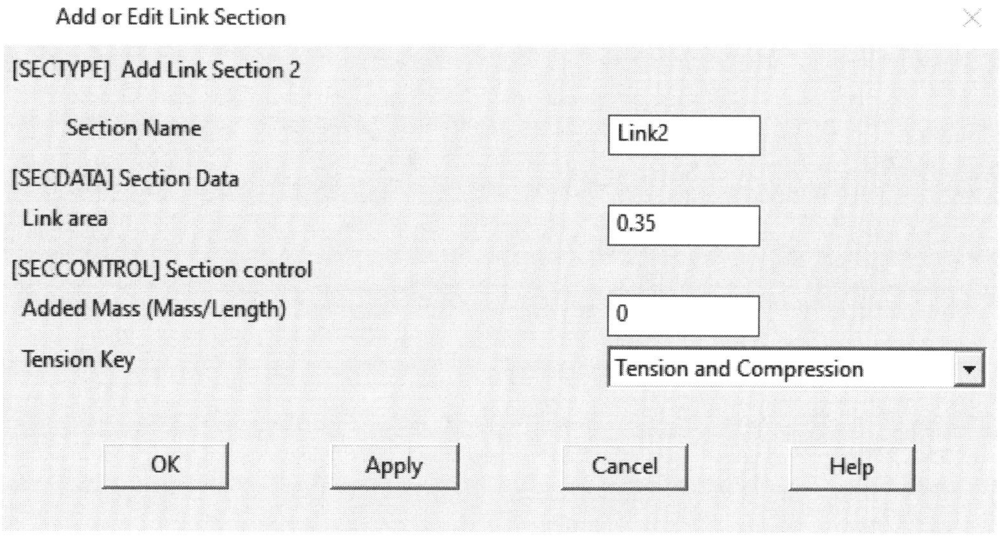

Add or Edit Link Section

[SECTYPE] Add Link Section 2

Section Name — Link2

[SECDATA] Section Data

Link area — 0.35

[SECCONTROL] Section control

Added Mass (Mass/Length) — 0

Tension Key — Tension and Compression

OK   Apply   Cancel   Help

Figure 1-26 Real constant set No. 2.

Close the Real Constants window which now shows Set 1 and Set 2 > **Close**

Enter the material properties next.

5. Main Menu > Preprocessor > Material Props > Material Models

Material Model Number 1

Click **Structural > Linear > Elastic > Isotropic**

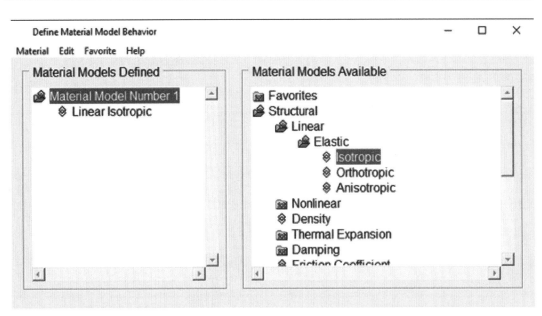

Figure 1-27 Material models.

Enter **EX = 3.E7** and **PRXY = 0.27 > OK**

Linear Isotropic Properties for Material Number 1   ✕

Linear Isotropic Material Properties for Material Number 1

	T1
Temperatures	0
EX	3e7
PRXY	0.27

Add Temperature	Delete Temperature		Graph
	OK	Cancel	Help

Figure 1-28 Properties for material no. 1.

For the Second material: In the Define Material Model Behavior window, **Material > New Model Define Material ID** > 2

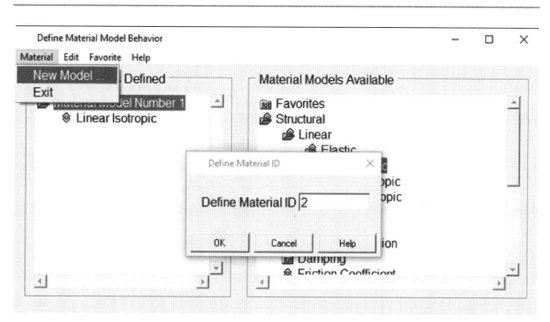

Figure 1-29 Define material ID no. 2.

Enter **EX = 1.2E7** and **PRXY = 0.3 > OK**

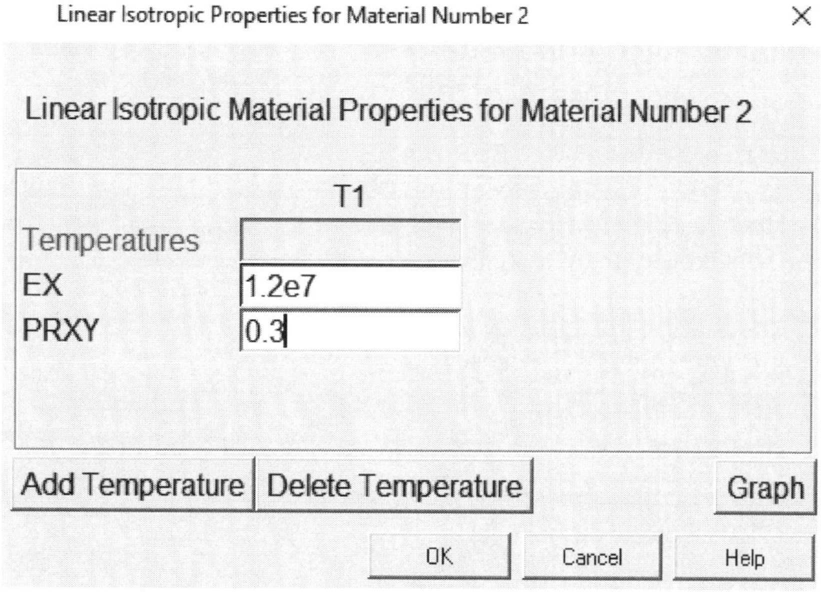

Linear Isotropic Properties for Material Number 2   ×

Linear Isotropic Material Properties for Material Number 2

	T1
Temperatures	
EX	1.2e7
PRXY	0.3

Add Temperature Delete Temperature   Graph

OK   Cancel   Help

Figure 1-30 Properties for material no. 2.

Close the Define Material Model Behavior window.

Now enter **nodal coordinate** data and **element connectivity** information.

6. Main Menu > Preprocessor > Modeling > Create > Nodes > In Active CS

For Node 1: Enter **Node number = 1; X = 0.0, Y = 0.0 > Apply**   (Note use of *Apply*)

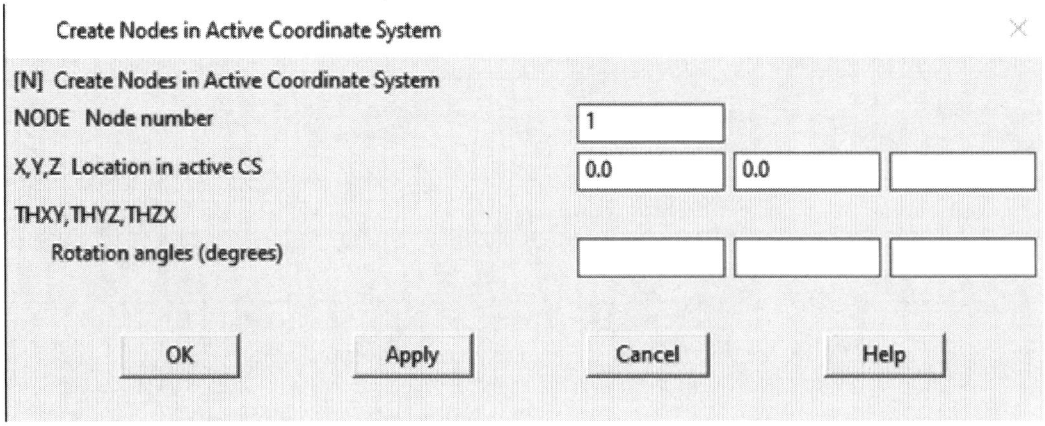

Create Nodes in Active Coordinate System   ✕

[N] Create Nodes in Active Coordinate System

NODE   Node number   [1]

X,Y,Z Location in active CS   [0.0]   [0.0]   [   ]

THXY,THYZ,THZX

        Rotation angles (degrees)   [   ]   [   ]   [   ]

        OK   Apply   Cancel   Help

Figure 1-31 Create nodes.

For Nodes 2 and 3:

Enter **Node number = 2; X = 20.0, Y = 0.0 > Apply**

Enter **Node number = 3; X = 0.0, Y = 15.0 > OK**

(CS stands for Coordinate System. Because this problem is a 2D, X-Y plane model, no Z-coordinate values need be entered. It does not hurt anything to add them if you wish, however.)

If you make a mistake, return to this Node Create window to correct the coordinate values. You can use **Preprocessor > Modeling > Delete** to remove nodes.

Next select Material Property (1) and Real Property Set (1), then create element (1).

7. Main Menu > Preprocessor > Modeling > Create > Elements > Elem Attributes

Select defaults, Material number **> 1** and Section number **> 1 > OK**

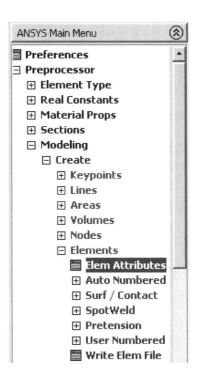

Figure 1-32 Select element attributes.

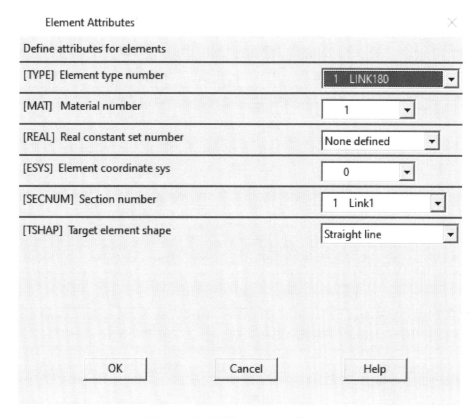

Figure 1-33 Element attributes.

8. Main Menu > Preprocessor > Modeling > Create > Elements > Auto Numbered > Thru Nodes

Pick **Node 1** then **Node 2 > OK**

Now select the second set of material and real properties, and create element (2).

9. Main Menu > Preprocessor > Modeling > Create > Elements > Elem Attributes

Select Material number > **2** and Real constant set number > **2 > OK**

10. Main Menu > Preprocessor > Modeling > Create > Elements > Auto Numbered > Thru Nodes

Pick **Node 2** then **Node 3 > OK**

In this simple example we have only two elements, and they each have different properties. If several elements have the same material and real properties, you only need to set the properties once, then create all those elements before changing to the next set of material and real properties. Now let's **check** the element properties to make sure the input was correct.

11. Utility Menu > List > Elements > Nodes+Attr + RealConst

```
LIST ALL SELECTED ELEMENTS.   (LIST NODES)

    ELEM MAT TYP REL ESY SEC          NODES

        1   1   1   1   0   1      1     2
     AREA          ADMS        TFLG                          CV1
CV2
    0.125000       0.00000      0.00000      0.00000      0.00000
0.00000
        2   2   1   2   0   1      2     3
     AREA          ADMS        TFLG                          CV1
CV2
    0.350000       0.00000      0.00000      0.00000      0.00000
0.00000
```

The element data is correct, so we can now apply the boundary conditions and loadings, both located under 'Loads' in ANSYS.

12. Main Menu > Preprocessor > Loads > Define Loads > Apply > Structural > Displacement > On Nodes

Pick **Node 1 > OK > All DOF > Constant value > Displacement value = 0 > Apply**

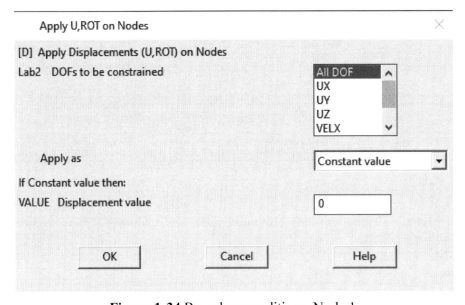

Figure 1-34 Boundary conditions, Node 1.

(A blank entry for value is interpreted as a zero. If you make a mistake, return to this window or use **Loads > Delete . . .** to make corrections.)

For Node 3:

Pick **Node 3 > OK > UX > Constant value** > Enter **–0.01 > Apply**

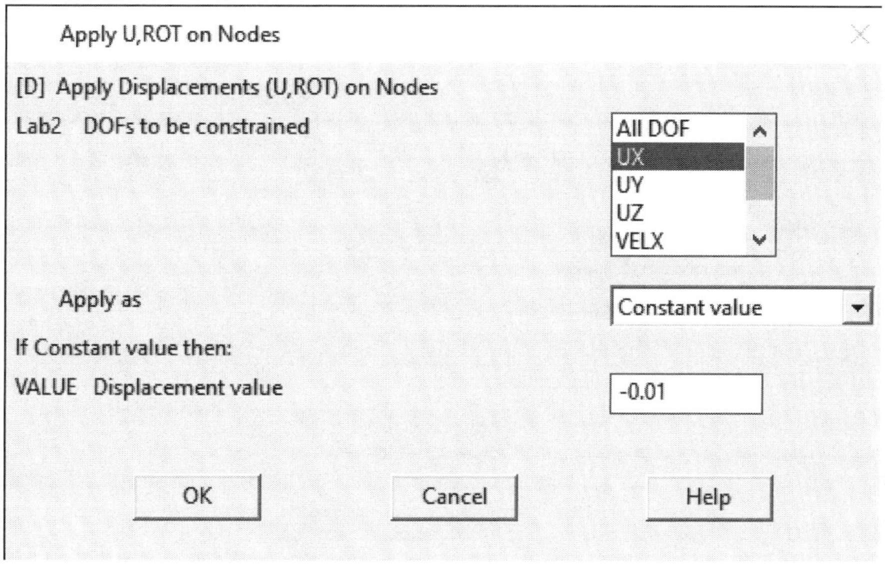

Figure 1-35 Boundary condition, node 3.

Again pick **Node 3 > OK > UY > Constant value** > Enter **0 > OK**

Finally we will apply the 300 lbf forces.

13. Main Menu > Preprocessor > Loads > Define Loads > Apply > Structural > Force/Moment > On Nodes

Pick Nodes **1** *and* **2, > OK.** Select **FY**, and **Constant value**, Enter **–300 > OK**

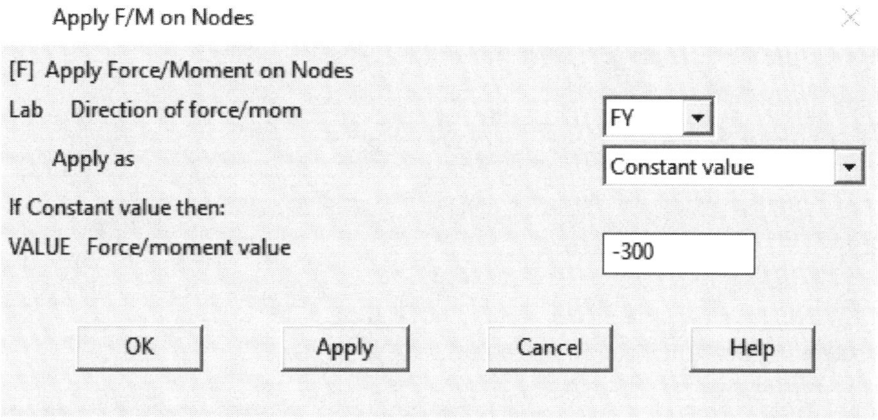

Figure 1-36 Loads.

Let's check the boundary conditions and loads. Use the **Utility Menu**, top of the screen.

14. Utility Menu > List > Loads > DOF Constraints > On All Nodes

```
DLIST Command                                                          X
File

LIST CONSTRAINTS FOR SELECTED NODES          1 TO         3 BY         1
CURRENTLY SELECTED DOF SET= UX    UY    UZ

    NODE  LABEL      REAL           IMAG
       1  UX      0.00000000     0.00000000
       1  UY      0.00000000     0.00000000
       1  UZ      0.00000000     0.00000000
       3  UX    -0.100000000E-01 0.00000000
```

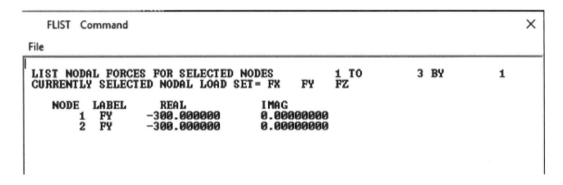

Figure 1-37 List constraints.

15. Utility Menu > List > Loads > Forces > On All Nodes

```
FLIST Command                                                          X
File

LIST NODAL FORCES FOR SELECTED NODES         1 TO         3 BY         1
CURRENTLY SELECTED NODAL LOAD SET= FX    FY    FZ

   NODE  LABEL      REAL          IMAG
      1  FY      -300.000000    0.00000000
      2  FY      -300.000000    0.00000000
```

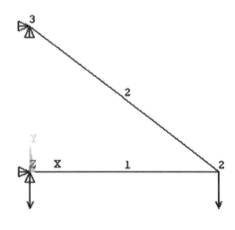

Figure 1-38 List nodal forces.

For a visual check: **Utility Menu > PlotCtrls** turn on **Node and Element Numbers** (Numbering . . .), and **All Applied BCs** (Symbols . . .)

16. Utility Menu > Plot > Elements

The preprocessing is complete and we can save the model **database**. Solve the problem as before.

Figure 1-39 Model plot.

17. Utility Menu > File > Save as Jobname.db   (or **Save as . . .** you supply a name.)

1-7 HOW IT WORKS

In finite element analysis or in matrix structural analysis, a matrix is developed to describe the input-output behavior of each element. In this lesson the elements are **truss elements**, and the behavior we need is the relationship between the nodal loads **f** and the nodal displacements **u**. This matrix is called the **element stiffness matrix**.

To analyze a structure composed of many truss elements, each element stiffness matrix is evaluated according to its length, area, and material modulus of elasticity. Next, a mathematical model of the entire structure is created by **assembling** the individual element stiffness matrices, much like one builds a mathematical model of an electrical circuit from electrical resistor, capacitor, etc. component models. At a node where two trusses join, the overall stiffness of the assembled structure (resistance to an external force) is increased by the sum of the stiffness of the individual elements that come together at that node. For **linear analysis**, the resulting global problem is the set of linear algebraic equations shown below.

$$\{P\} = [K]\,\{U\}$$

For the problem in Tutorial 1A, {**P**} is the vector of external loads **Fx** and **Fy**, [**K**] is the global stiffness matrix assembled from the element stiffness matrices, and {**U**} is the vector of node point displacements **Ux** and **Uy**. {**P**} and [**K**] are known from the definition of the problem under consideration, and {**U**} is to be found by solving the matrix equation above. If the structure can move as a **rigid body** when a force is applied, [**K**] will be **singular**, and the equations **cannot be solved**.

An important part of finite element analysis is getting the boundary conditions right. For static loads analysis, the object must have enough of the right kind of displacement constraints to **prevent rigid body motion** as mentioned previously. ANSYS provides an error message if the model is **under constrained** because of the singular stiffness matrix. In such a case the solution process halts. Select the item on the task bar line to examine the ANSYS Output window.

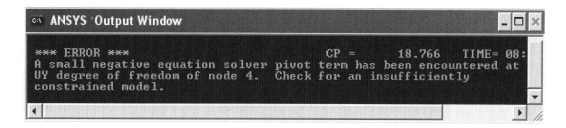

Figure 1-40 Model plot.

On the other hand, the model must also be **properly** constrained so it does not have boundary conditions that define a problem other than the one you really want to solve.

1-8 SUMMARY

Simple 2-D truss models were used in this lesson to introduce the general form of typical FEM models, to introduce basic ANSYS operations, to demonstrate both file-based and interactive analysis options, and to introduce some facets of the important engineering modeling decisions that must be considered in finite element analysis. Explanations of the use of gravity loadings and multiple load cases are presented in Lesson 8.

In Lesson 2 we consider problems in Plane Stress and Plane Strain wherein models of solid parts are developed using two-dimensional triangular or quadrilateral elements developed for that purpose. Modeling of this nature gives us the ability to consider complex geometries, boundary conditions and loadings that may occur in everyday engineering practice.

1-9 PROBLEMS

You can use the start-up dialog box or **File > Change Title** to add the problem number and your name to the title that appears on all plots associated with a particular job.

1-1 The truss shown has a horizontal joint spacing of 9 ft and a vertical joint spacing of 6 ft. It is made of steel pin-jointed tubes whose cross sections are 6.625 in outer diameter with a 0.28 wall thickness. Find the maximum displacement, the maximum stress and determine if any of the members in compression are likely to buckle. The vertical force is 50,000 lbf.

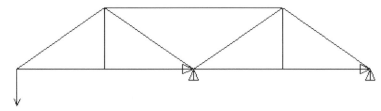

Figure P1-1

1-2 Solve problem 1-1 again but change the horizontal and vertical elements to aluminum tubes with an 8.625 inch OD and a 0.322 inch wall thickness.

1-3 The truss shown is made of steel from pin-jointed elements that have a square cross section 15 mm on a side. The horizontal joint spacing is 2 m. The vertical spacing is 4 m and 2 m. The downward load is 3000 N. Find the maximum displacement, the highest stressed element, and evaluate the possibility of local buckling.

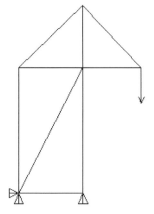

Figure P1-3

1-4 Solve problem 1-3 if the three horizontal members are made of aluminum with a 20 mm x 20 mm square cross section.

1-5 Find the magnitude and location of the maximum displacement and the maximum stress in the 2-D model of a pin jointed tower shown below. All elements are 4 x 4 x 3/8 steel angle sections with area 2.86 sq in. The load **P** has a value equal to 4500 lbf.

Are any elements likely to buckle if the section minimum flexural inertia is 2.1 in⁴?

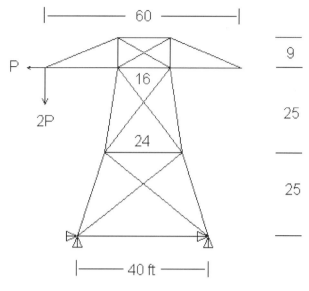

Figure P1-5

1-6 The truss in figure P1-6 is made from steel elements with 2 inch diameter solid circular cross sections. The horizontal spacing is 40 ft, the vertical spacing from the supports up is 60 ft., 40 ft., and 30 ft. A downward load of 1000 lbf is applied at each of the upper nodes. Find the maximum stress and deflection.

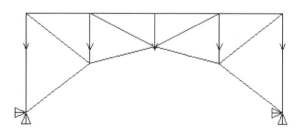

Figure P1-6

NOTE: In recent versions, the 'real constant' notation used above has been replaced by 'sectype' and 'secdata' as shown below. ANSYS APDL currently accepts either form.

```
et, 1, link180          ! Element type; no.1 is link180
sectype, 1, link        ! Type of cross section is link
secdata, 0.5            ! Cross sectional area = 0.5 sq in
```

NOTES:

Lesson 2

Plane Stress
Plane Strain

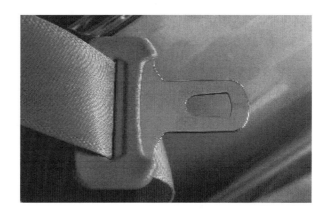

2-1 OVERVIEW

Plane stress and plane strain problems are an important subclass of general three-dimensional stress and strain problems. The tutorials in this lesson demonstrate:

♦ Solving planar stress concentration problems.

♦ Evaluating potential inaccuracies in the solutions.

♦ Using the various ANSYS 2D element formulations.

2-2 INTRODUCTION

It is possible for an object such as the one on the cover of this book to have six components of stress when subjected to arbitrary three-dimensional loadings. When referenced to a Cartesian x, y, z coordinate system these components of stress are:

Normal Stresses $\sigma_x,\ \sigma_y,\ \sigma_z$

Shear Stresses $\tau_{xy},\ \tau_{yz},\ \tau_{zx}$

Figure 2-1 Stresses in 3 dimensions.

In general, the analysis of such objects requires three-dimensional modeling as discussed in Lesson 4. However, two-dimensional models are often easier to develop, easier to solve and can be employed in many situations **if** they can accurately represent the behavior of the object under loading.

A state of **Plane Stress** exists in a thin object loaded in the plane of its largest dimensions. Let the *X-Y* plane be the plane of analysis. The non-zero stresses σ_x, σ_y, and τ_{xy} lie in the *X* - *Y* plane and do not vary in the *Z* direction. Further, the other stresses (σ_z, τ_{yz} , and τ_{zx}) are all zero for this kind of geometry and loading. A thin beam loaded in its plane and a spur gear tooth are good examples of plane stress problems.

ANSYS provides a 6-node planar triangular element along with 4-node and 8-node quadrilateral elements for use in the development of plane stress models. We will use both triangles and quads in solution of the example problems that follow.

2-3 PLATE WITH CENTRAL HOLE

To start off, let's solve a problem with a known solution so that we can check our computed results as well as our understanding of the FEM process. The problem is that of a tensile-loaded thin plate with a central hole as shown in Figure 2-2.

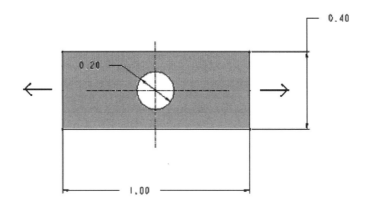

Figure 2-2 Plate of 0.01 meter thickness with a central hole.

The **1.0 m x 0.4 m** plate has a **thickness** of **0.01 m**, and a central hole **0.2 m** in **diameter**. It is made of steel with material properties; **elastic modulus**, *E* = **2.07 x 10¹¹ N/m²** and **Poisson's ratio**, *ν* = **0.29**. We apply a **horizontal tensile loading** in the form of a **pressure *p*** = **-1.0 N/m²** along the vertical edges of the plate. Entering data with consistent units is the engineer's responsibility.

Because holes are necessary for fasteners such as bolts, rivets, etc, the need to know stresses and deformations near them occur very often and have received a great deal of study. The results of these studies are widely published, and we can look up the stress concentration factor for the case shown above. Before the advent of suitable computation methods, the effect of most complex stress concentration geometries had to be evaluated experimentally, and many available charts were developed from experimental results.

The uniform, homogeneous plate above is symmetric about horizontal axes in both geometry and loading. This means that the state of stress and deformation below a horizontal centerline is a mirror image of that above the centerline, and likewise for a vertical centerline. We can take advantage of the symmetry and, by applying the correct boundary conditions, use only a quarter of the plate for the finite element model. For small problems using symmetry may not be too important; for large problems it can save modeling and solution efforts by eliminating one-half or a quarter or more of the work.

Place the origin of X-Y coordinates at the center of the hole. If we pull on both ends of the plate, points on the centerlines will move along the centerlines but not perpendicular to them. This indicates the appropriate displacement conditions to use as shown below.

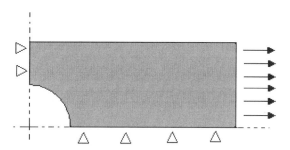

Figure 2-3 Quadrant used for analysis.

In Tutorial 2A we will use ANSYS to determine the maximum horizontal stress in the plate and compare the computed results with the maximum value that can be calculated using tabulated values for stress concentration factors. Interactive commands will be used to formulate and solve the problem.

2-4 TUTORIAL 2A - PLATE

Objective: Find the **maximum axial stress** in the plate with a central hole and compare your result with a computation using published stress concentration factor data.

PREPROCESSING

1. Start ANSYS, select the **Working Directory** where you will store the files associated with this problem. Also set the **Jobname** to **Tutorial2A** or something memorable and provide a **Title**.

(If you want to make changes in the Jobname, working Directory, or Title after you've started ANSYS, use **File > Change Jobname** or **Directory** or **Title**.)

Select the **six node triangular element** to use for the solution of this problem.

Figure 2-4 Six-node triangle.

The six-node triangle is a **sub-element** of the eight-node quadrilateral.

2. Main Menu > Preprocessor > Element Type > Add/Edit/Delete > Add > Structural Solid > Quad 8node 183 > OK

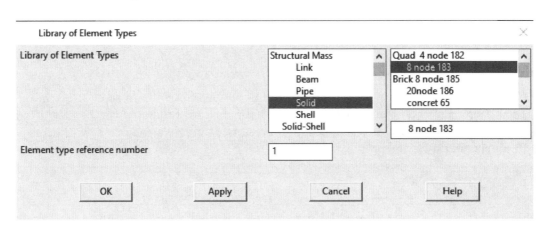

Figure 2-5 Element selection.

Select the triangle **option** and the option to define the plate thickness, otherwise a unit thickness is used.

3. Options (Element shape K1) > **Triangle,**

      Options (Element behavior K3) > **Plane strs w/thk > OK > Close**

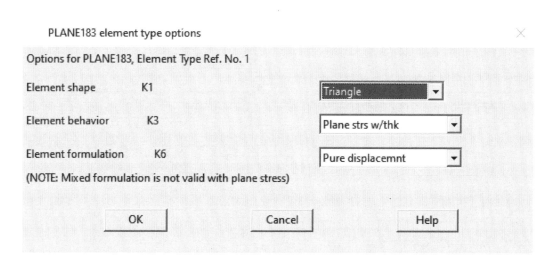

Figure 2-6 Element options.

4. Main Menu > Preprocessor > Real Constants > Add/Edit/Delete > Add > OK

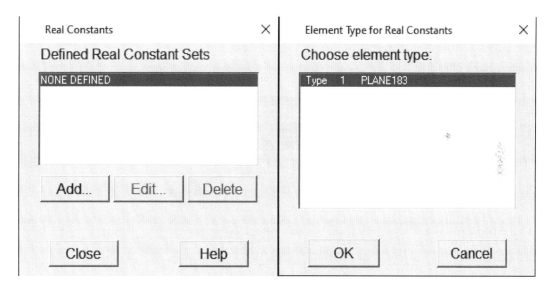

Figure 2-7 Real constants.

(Enter the plate thickness of 0.01 m.) >Enter **0.01 > OK > Close**

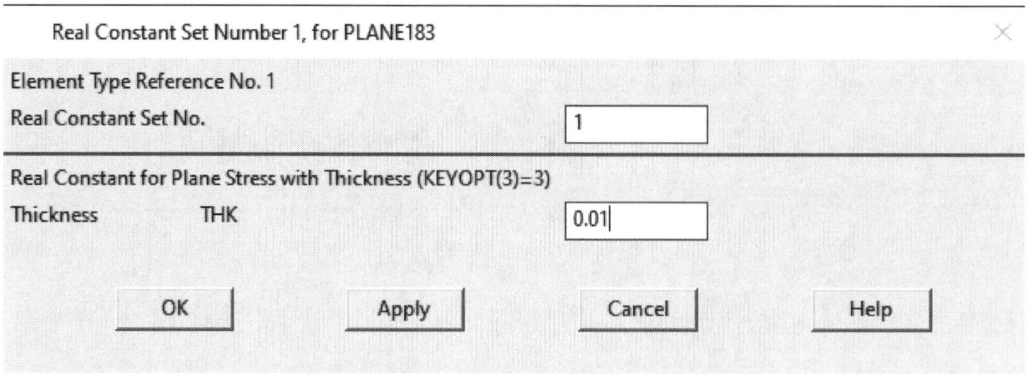

Real Constant Set Number 1, for PLANE183   ✕

Element Type Reference No. 1
Real Constant Set No.   [1]

Real Constant for Plane Stress with Thickness (KEYOPT(3)=3)
Thickness   THK   [0.01|]

        OK   Apply   Cancel   Help

Figure 2-8 Enter the plate thickness.

Enter the material properties.

5. Main Menu > Preprocessor > Material Props > Material Models

Material Model Number 1, click **Structural > Linear > Elastic > Isotropic**

Enter **EX = 2.07E11** and **PRXY = 0.29 > OK** (**Close** the Define Material Model Behavior window.)

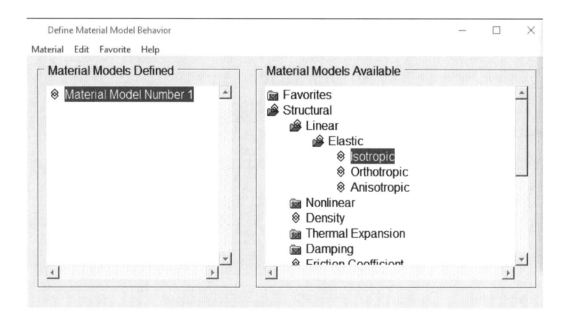

Figure 2-9 Material Models.

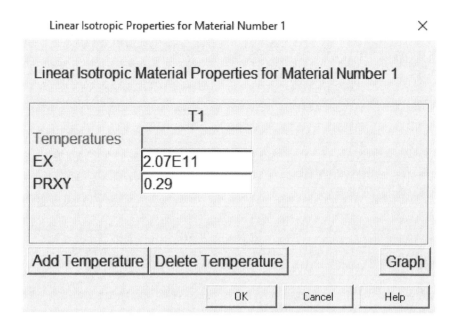

Figure 2-10 Material Properties.

Create the geometry for the upper right quadrant of the plate by **subtracting** a 0.2 m diameter circle from a 0.5 x 0.2 m rectangle. Generate the rectangle first.

6. Main Menu > Preprocessor > Modeling > Create > Areas > Rectangle > By 2 Corners

Enter (lower left corner) **WP X = 0.0, WP Y = 0.0** and **Width = 0.5, Height = 0.2 > OK**

7. Main Menu > Preprocessor > Modeling > Create > Areas > Circle > Solid Circle

Enter **WP X = 0.0, WP Y = 0.0** and **Radius = 0.1 > OK**

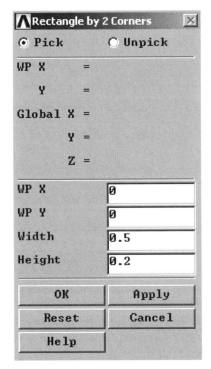
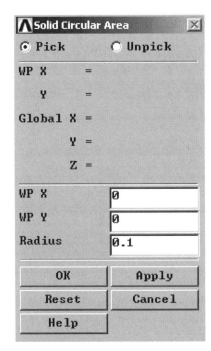

Figure 2-11 Create areas.

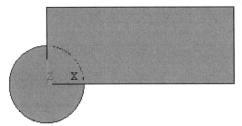

Figure 2-12 Rectangle and circle.

Now **subtract** the circle from the rectangle. (Read the messages in the window at the bottom of the screen as necessary.) Use (top of screen) Raise Hidden, Reset Picking as necessary.

8. Main Menu > Preprocessor > Modeling > Operate > Booleans > Subtract > Areas > Pick the rectangle > OK, then pick the circle **> OK** (Use **Raise Hidden** and **Reset Picking** as necessary.)

Figure 2-13 Geometry for quadrant of plate.

Create a mesh of triangular elements over the quadrant area.

9. Main Menu > Preprocessor > Meshing > Mesh > Areas > Free Pick the quadrant >
OK   (Note that your mesh may differ from that shown here.)

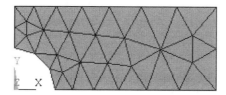

Figure 2-14 Triangular element mesh.

Apply the displacement boundary conditions and loads to the **geometry (lines)** instead of
the nodes as we did in the previous lesson. These conditions will be applied to the FEM
model when the solution is performed.

**10. Main Menu > Preprocessor > Loads > Define Loads > Apply > Structural >
Displacement > On Lines** Pick the left edge of the quadrant > **OK > UX = 0. > OK**

**11. Main Menu > Preprocessor > Loads > Define Loads > Apply > Structural >
Displacement > On Lines** Pick the bottom edge of the quadrant > **OK > UY = 0. > OK**

Apply the loading.

**12. Main Menu > Preprocessor > Loads > Define Loads > Apply > Structural >
Pressure > On Lines.** Pick the right edge of the quadrant > **OK > Pressure = -1.0 > OK**
(A positive pressure would be a compressive load, so we use a *negative* pressure. The
pressure is shown by the two arrows.)

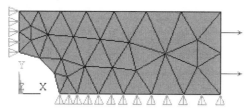

Figure 2-15 Model with loading and displacement boundary conditions.

The model-building step is now complete, and we can proceed to the solution. First, to be
safe, save the model.

13. Utility Menu > File > Save as Jobname.db   (Or **Save as ....** ; use a new name)

SOLUTION

The interactive solution proceeds as illustrated in the tutorials of Lesson 1.

14. Main Menu > Solution > Solve > Current LS > OK

The **/STATUS Command** window displays the problem parameters and the **Solve Current Load Step** window is shown. Check the solution options in the /STATUS window and if all is OK, select **File > Close**

In the **Solve Current Load Step** window, select **OK**, and when the solution is complete, **Close** the '**Solution is Done!**' window.

POSTPROCESSING

We can now **plot the results** of this analysis and also **list the computed values**. First examine the **deformed shape. PlotCtrls > Symbols > All Applied BCs**

15. Main Menu > General Postproc > Plot Results > Deformed Shape > Def. + Undef. > OK

```
DISPLACEMENT
STEP=1
SUB =1
TIME=1
DMX =.320E-11

U
```

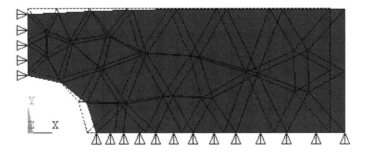

Figure 2-16 Plot of Deformed shape.

The deformed shape looks correct. (The undeformed shape is indicated by the dashed lines.) The right end moves to the right in response to the tensile load in the *X* direction, the circular hole ovals out, and the top moves down because of Poisson's effect. Note that the element edges on the circular arc are represented by straight lines. This is an artifact of the plotting routine not the analysis. The six-node triangle has curved sides, and if you plot the nodes (Plot > Nodes), you will see that a node is placed on the curved edges.

The maximum displacement is shown on the graph legend as 0.32e-11 which seems reasonable. The **units** of displacement are **meters** because we employed meters and N/m² in the problem formulation. Now read the data and plot the stress in the X direction.

Main Menu > General Postproc > Read Results > First Set

Use **PlotCtrls > Symbols [/PSF] Surface Load Symbols** (set to **Pressures**) and **Show pre and convect as** (set to **Arrows**) to display the pressure loads.

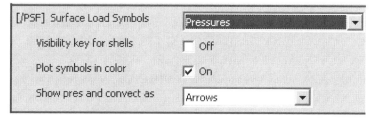

Figure 2-17 Surface load symbols.

Also select **Display All Applied BCs**

16. **Main Menu > General Postproc > Plot Results > Contour Plot > Element Solu > Stress > X-Component of stress > OK**

NOTE: We are plotting the **ELEMENT SOLUTION**

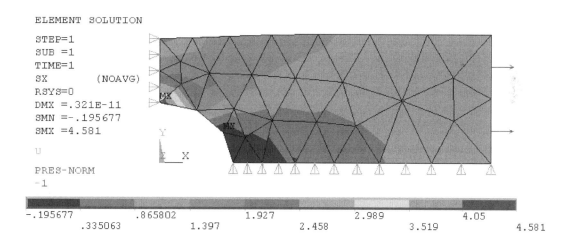

Figure 2-18 Element SX stresses.

The minimum, **SMN,** and maximum, **SMX,** stresses as well as the color bar legend give an overall evaluation of the σ_x (SX) stress state. We are interested in the maximum stress at the hole. Use the **Zoom** to focus on the area with highest stress. (Your meshes and results may differ a bit from those shown here.)

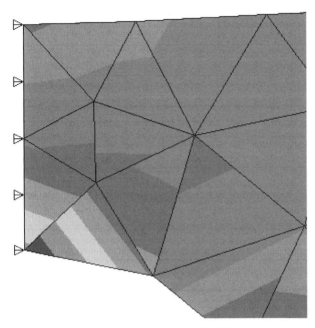

Figure 2-19 SX stress detail.

Stress variations in the actual isotropic, homogeneous plate should be smooth and continuous across elements. The **discontinuities** in the SX stress contours above indicate that the number of elements used in this model is too few to calculate with complete accuracy the stress values near the hole because of the stress gradients there. We **will not accept** this stress solution. More six-node elements are needed in the region near the hole to find accurate values of the stress. On the other hand, in the right half of the model, away from the stress riser, the calculated stress contours are smooth, and SX would seem to be accurately determined there.

It is **important** to note that in the plotting we selected **Element Solu** (Element Solution) in order to look for stress contour discontinuities. If you pick **Nodal Solu** to plot instead, for problems like the one in this tutorial, the stress values will be **averaged** before plotting, and any contour discontinuities (and thus **errors**) will be hidden. If you plot nodal solution stresses you will always see smooth contours.

A word about **element accuracy:** The FEM implementation of the **truss** element is taken directly from solid mechanics studies, and there is no approximation in the solutions for node-loaded truss structures formulated and solved in the ways discussed in Lesson 1.

The **continuum elements** such as the ones for plane stress and plane strain, on the other hand, are normally developed using displacement functions of a polynomial type to represent the displacements within the element. The higher the polynomial, the greater the accuracy. The ANSYS six-node triangle uses a quadratic polynomial and is capable of representing **linear stress and strain variations** within an element. Note the SX plot.

Near stress concentrations the stress gradients vary quite sharply. To capture this variation, the number of elements near the stress concentrations must be increased proportionately.

To obtain more elements in the model, return to the Preprocessor and refine the mesh, first remove the pressure. All elements are subdivided and the mesh below is created.

17. Main Menu > Preprocessor > Loads > Define Loads > Delete > Structural > Pressure > On Lines. Pick the right edge of the quadrant. **Main Menu > Preprocessor > Meshing > Modify Mesh > Refine At > All** (Select **Level of refinement 1.**) **> OK**

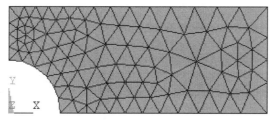

Figure 2-20 Global mesh refinement.

We will also refine the mesh selectively near the hole.

18. Main Menu > Preprocessor > Meshing > Modify Mesh > Refine At > Nodes. (Select the three nodes shown.) **> OK** (Select the **Level of refinement = 1**) **> OK**

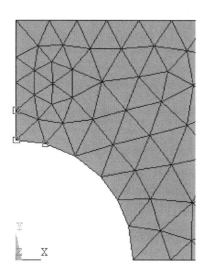 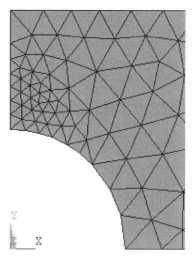

Figure 2-21 Selective refinement at nodes.

(Note: Alternatively you can use **Preprocessor > Meshing > Clear > Areas** to remove all elements and build a completely new mesh. **Plot > Areas** afterwards to view the area again. Note also that too much local refinement can create a mesh with too rapid a transition between fine and coarse mesh regions.)

Reapply the pressure loading, repeat the solution, and replot the stress SX.

19. Main Menu > Solution > Solve > Current LS > OK

Save your work.

20. File > Save as Jobname.db (Or **Save as ....** ; use a new name)

Main Menu > General Postproc > Read Results > First Set

21. Main Menu > General Postproc > Plot Results > Contour Plot > Element Solu > Stress > X-Component of stress > OK

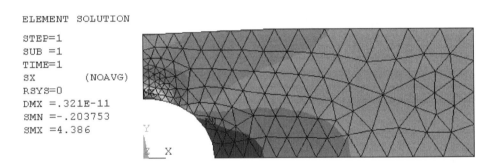

Figure 2-22 SX stress contour after mesh refinement.

Figure 2-23 SX stress detail contour after mesh refinement.

The element solution stress contours are now smooth across element boundaries, and the stress legend shows a maximum value of **4.386 *Pa***, a 4.3 percent change in the SX stress computed using the previous mesh.

To check this result, find the stress concentration factor for this problem in a text or reference book or from a suitable web site. For the geometry of this example we find $K_t = 2.17$. We can compute the maximum stress using (K_t)(load)/(net cross sectional area). Using the pressure $p = 1.0$ *Pa* we obtain. (Load = pressure x area.)

$$\sigma_{x\,MAX} = 2.17 * p * (0.4)(0.01)/[(0.4 - 0.2) * 0.01] = 4.34 Pa$$

The computed maximum value is **4.39 *Pa*** which is around **one percent in error**, *assuming* that the value of K_t is exact.

2-5 THE APPROXIMATE NATURE OF FEM

As mentioned above, the stiffness matrix for the truss elements of Lesson 1 can be developed directly and simply from elementary solid mechanics principles. For continuum problems in two and three-dimensional stress, this is generally no longer possible, and the element stiffness matrices are usually developed by assuming something specific about the characteristics of the displacements that can occur within an element.

Ordinarily this is done by specifying the highest degree of the polynomial that governs the displacement distribution within an element. For **h-method** elements, the polynomial degree depends upon the **number of nodes** used to describe the element, and the interpolation functions that relate displacements within the element to the displacements at the nodes are called **shape functions**. In ANSYS, 2-dimensional problems can be modeled with **six-node triangles**, **four-node quadrilaterals** or **eight-node quadrilaterals**.

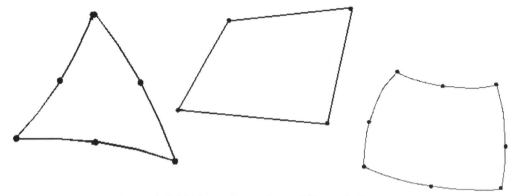

Figure 2-24 Triangular and quadrilateral elements.

The greater the number of nodes, the higher the order of the polynomial and the greater the accuracy in describing displacements, stresses and strains within the element. If the stress is constant throughout a region, a very simple model is sufficient to describe the stress state, perhaps only one or two elements. If there are gradients in the stress distributions within a region, high-degree displacement polynomials and/or many elements are required to accurately analyze the situation.

These comments explain the variation in the accuracy of the results as different numbers of elements were used to solve the problem in the previous tutorial and why the engineer must carefully prepare a model, start with small models, grow the models as understanding of the problem develops and carefully interpret the calculated results. The ease with which models can be prepared and solved sometimes leads to careless evaluation of the computed results.

2-6 ANSYS FILES

The files created during the solution were saved in step 20 of Tutorial 2A. Look in the working directory and you see Tutorial2A files with extensions **BCS**, **db**, **dbb**, **esav**, **full**, **mntr**, **rst**, and **stat**. However, the Tutorial 2A problem can be reloaded using only **Tutorial2A.db**, so if you want to save disk space, you can delete the others.

2-7 ANSYS GEOMETRY

The finite element model consists of elements and nodes and is separate from the geometry on which it may be based. It is possible to build the finite element model without consideration of any underlying geometry as was done in the truss examples of Lesson 1, but in many cases, development of the geometry is the first task.

Two-dimensional geometry in ANSYS is built from **keypoints**, **lines** (straight, arcs, splines), and **areas**. These geometric items are assigned numbers and can be listed, numbered, manipulated, and plotted. The keypoints (2,3,4,5,6), lines (2,3,5,9,10), and area (3) for Tutorial 2A are shown below. (Your numbering may differ.)

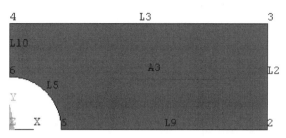

Figure 2-25 Keypoints, lines and areas.

The finite element model developed previously for this part used the area A3 for development of the node/element FEM mesh. The loads, displacement boundary conditions and pressures were applied to the geometry lines. When the solution step was executed, the loads were transferred from the lines to the FEM model nodes. Applying boundary conditions and loads to the geometry facilitates remeshing the problem. The geometry does not change, only the number and location of nodes and elements, and at solution time, the loads are transferred to the new mesh.

Geometry can be created in ANSYS interactively (as was done in the previous tutorial) or it can be created by reading a text file. For example, the geometry of Tutorial 2A can be generated with the following text file using the **File > Read Input from** command sequence. (The keypoint, line, etc. numbers will be different from those shown above.)

```
/FILNAM,Geom
/title, Stress Concentration Geometry
! Example of creating geometry using keypoints, lines, arcs
/prep7
! Create geometry
k, 1,  0.0,  0.0      ! Keypoint 1 is at 0.0, 0.0
k, 2,  0.1,  0.0
k, 3,  0.5,  0.0
k, 4,  0.5,  0.2
k, 5,  0.0,  0.2
k, 6,  0.0,  0.1

L, 2, 3               ! Line from keypoints 2 to 3
L, 3, 4
L, 4, 5
L, 5, 6

! arc from keypoint 2 to 6, center kp 1, radius 0.1
LARC, 2, 6, 1, 0.1

AL, 1, 2, 3, 4, 5    ! Area defined by lines 1,2,3,4,5
```

Geometry for FEM analysis also can be created with solid modeling **CAD** or other software and imported into ANSYS. The **IGES** (Initial Graphics Exchange Specification) neutral file is a common format used to exchange geometry between computer programs. Tutorial 2B demonstrates this option for ANSYS geometry development.

2-8 TUTORIAL 2B – SEATBELT COMPONENT

Objective: Determine the stresses and deformation of the prototype seatbelt component shown in the figure below if it is subjected to tensile load of **1000 lbf**.

Figure 2-26 Seatbelt component.

The seatbelt component is made of steel, has an over all length of about 2.5 inches and is 3/32 = 0.09375 inches thick. A solid model of the part was developed in a CAD system and exported as an IGES file. The file is imported into ANSYS for analysis. For simplicity we will analyze only the right, or 'tongue' portion of the part in this tutorial.

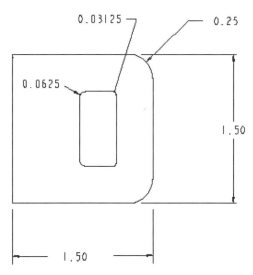

Figure 2-27 Seatbelt 'tongue'.

PREPROCESSING

1. Start ANSYS, Run Interactive, set jobname, and working directory.

Create the top half of the geometry above. The latch retention slot is 0.375 x 0.8125 inches and is located 0.375 inch from the right edge.

If you are not using an IGES file to define the geometry for this exercise, you can create the geometry directly in ANSYS with key points, lines, and arcs by selecting **File > Read Input from** to read in the text file given below and by **skipping the IGES import steps 2, 3, 4,** and **10 below**.

```
/FILNAM,Seatbelt
/title, Seatbelt Geometry
! Example of creating geometry using keypoints, lines, arcs
/prep7
! Create geometry
k, 1,   0.0,     0.0         ! Keypoint 1 is at 0.0, 0.0
k, 2,   0.75,    0.0
k, 3,   1.125,   0.0
k, 4,   1.5,     0.0
k, 5,   1.5,     0.5
k, 6,   1.25,    0.75
k, 7,   0.0,     0.75
k, 8,   1.125,   0.375
k, 9,   1.09375, 0.40625
k, 10,  0.8125,  0.40625
k, 11,  0.75,    0.34375
k, 12,  1.25,    0.5
k, 13,  1.09375, 0.375
k, 14,  0.8125,  0.34375
```

```
L,  1,  2                        ! Line from keypoints 1 to 2
L,  3,  4
L,  4,  5
L,  6,  7
L,  7,  1
L,  3,  8
L,  9,  10
L,  11,  2

! arc from keypoint 5 to 6, center kp 12, radius 0.25, etc.
LARC,  5,6,  12,  0.25
LARC,  8,  9,  13,  0.03125
LARC,  10,  11,  14,  0.0625

AL,all                          ! Use all lines to create the area.
```

2. Alternatively, use a solid modeler to create the top half of the component shown above in the X-Y plane and export an IGES file of the part.

To import the IGES file

3. Utility Menu > File > Import > IGES > OK

Select the IGES file you created earlier. Accept the ANSYS import default settings. Merging is used to remove inconsistencies that may exist in the IGES file, for example keypoints that, because of the modeling or the file translation process, do not quite join to digital precision accuracy.

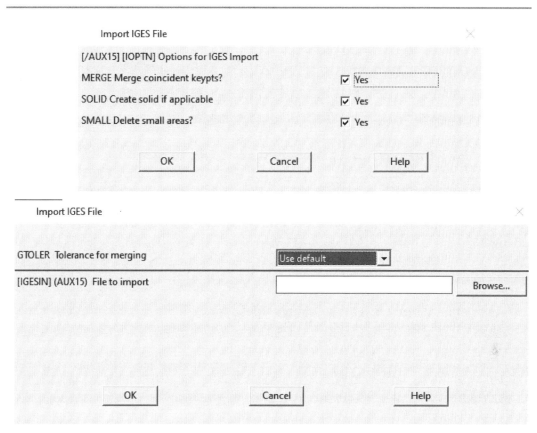

Figure 2-28 IGES import.

Turn the IGES solid model around if necessary so you can easily select the X-Y plane.
4. Utility Menu > PlotCtrls > Pan, Zoom, Rotate > Back, or use the side-bar icon.

Figure 2-29 Seatbelt solid, front and back.

5. Main Menu > Preprocessor > Element Type > Add/Edit/Delete > Add > Solid > Quad 8node   183 > OK (Use the 8-node quadrilateral element for this problem.)

6. Options > Plane strs w/thk > OK > Close

Enter the thickness

7. Main Menu > Preprocessor > Real Constants > Add/Edit/Delete > Add > (Type 1 Plane 183) **> OK >** Enter **0.09375 > OK > Close**

Enter the material properties

8. Main Menu > Preprocessor > Material Props > Material Models

Material Model Number 1, click **Structural > Linear > Elastic > Isotropic**

Enter **EX = 3.0E7** and **PRXY = 0.3 > OK** (Close Define Material Model Behavior window.)

Now mesh the X-Y plane area. (Turn on area numbers if it helps.)

9. Main Menu > Preprocessor > Meshing > Mesh > Areas > Free. Pick the X-Y planar area **> OK**

IMPORTANT NOTE: The mesh below was developed from an **IGES** geometry file. Using the text file geometry definition above may produce **a much different mesh**. If so, use the **Modify Mesh** refinement tools to obtain a mesh density that produces results with accuracies comparable to those given below. Computed **stress values** can be surprisingly **sensitive** to mesh differences. Smooth meshes can also be obtained by using six node triangular elements instead of eight node quads.

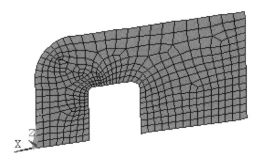

Figure 2-30 Quad 8 mesh.

The IGES **solid model** is no longer needed, and since its lines and areas may interfere with subsequent modeling operations, we can delete it from the session.

10. Main Menu > Preprocessor > Modeling > Delete > Volume and Below (Don't be surprised if everything disappears. Just **Plot > Elements** to see the mesh again.)

11. Utility Menu > PlotCtrls > Pan, Zoom, Rotate > Front (If necessary to see the front side of mesh.)

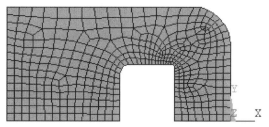

Figure 2-31 .Mesh, front view.

Now apply displacement and pressure boundary conditions. Zero displacement UX along left edge and zero UY along bottom edge.

12. Main Menu > Preprocessor > Loads > Define Loads > Apply > Structural > Displacement > On Lines Pick the left edge > **UX = 0.** > **OK**

13. Main Menu > Preprocessor > Loads > Define Loads > Apply > Structural > Displacement > On Lines Pick the lower edge > **UY = 0.** > **OK**

The 1000 lbf load corresponds to a uniform pressure of about 14,000 psi along the ¾ inch vertical inside edge of the latch retention slot. [1000 lbf/(0.09375 in. x 0.75 in.)].

14. Main Menu > Preprocessor > Loads > Define Loads > Apply > Structural > Pressure > On Lines

Select the inside line and set **pressure = 14000** > **OK**   (save your work)

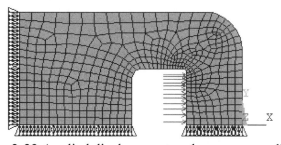

Figure 2-32 Applied displacement and pressure conditions.

Solve the equations.

SOLUTION

15. Main Menu > Solution > Solve > Current LS > OK

POSTPROCESSING

Comparing the **von Mises stress** with the **material yield stress** is an accepted way of evaluating static load yielding for **ductile metals** in a combined stress state, so we enter the postprocessor and plot the **element solution** of von Mises stress, **SEQV**.

16. Main Menu > General Postproc > Read Results > First Set

17. Main Menu > General Postproc > Plot Results > Contour Plot > Element Solu > Stress > (scroll down) von Mises > OK

Zoom in on the small fillet where the maximum stresses occur. The element solution stress contours are reasonably smooth, and the maximum von Mises stress is around **140,000 psi**. The small fillet radius of this geometry illustrates the challenges that can arise in creating accurate solutions; however, you can easily come within a few percent of the most likely true result using the methods discussed thus far.

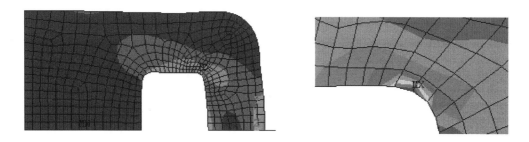

Figure 2-33 Von Mises stresses.

Redesign to reduce the maximum stress requires an increase in the thickness or fillet radius. Look at charts of stress concentration factors, and you notice that the **maximum stress increases** as the **radius of the stress raiser decreases**, approaching **infinite values at zero radii**.

If your model has a **zero radius notch**, your finite-size elements will show a very high stress but not infinite stress. If you refine the mesh, the stress will increase but not reach infinity. The finite element technique necessarily describes finite quantities and cannot directly treat an infinite stress at a **singular point**, so don't 'chase a singularity'. If you do not care what happens at the notch (static load, ductile material, etc.) do not worry about this location but examine the stresses and strains in other regions.

If you really are concerned about the maximum stress in a particular location (fatigue loads or brittle material), then use the actual part notch radius however small (1/32 for this tutorial); **do not use a zero radius**. Also examine the stress gradient in the vicinity of the notch to make sure the mesh is sufficiently refined near the notch. If a crack tip is the object of the analysis, you should look at **fracture mechanics** approaches to the problem. (See ANSYS help topics on fracture mechanics.)

The engineer's responsibility is not only to build useful models, but also to interpret the results of such models in intelligent and meaningful ways. This can often get overlooked in the rush to get answers.

Continue with the evaluation and check the strains and deflections for this model as well.

18. Main Menu > General Postproc > Plot Results > Contour Plot > Element Solu > Strain-total > 1st prin > OK

The maximum principal normal strain value is found to be approximately **0.004 in/in**.

19. Main Menu > General Postproc > Plot Results > Contour Plot > Nodal Solu > DOF Solution > X-Component of displacement > OK

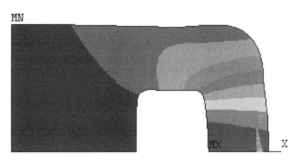

Figure 2-34 UX displacements.

The maximum deflection in the X direction is about **0.00145 inches** and occurs as expected at the center of the right-hand edge of the latch retention slot.

2-9 MAPPED MESHING

Quadrilateral meshes can also be created by mapping a square with a regular array of cells onto a general quadrilateral or triangular region. To illustrate this, **delete the last line,** AL,all, **from the text file** above so that the **area is not created** (just the lines) and read it into ANSYS. Use **PlotCtrls** to turn **Keypoint Numbering On**. Then use

1. Main Menu > Preprocessor > Modeling > Create > Lines > Lines > Straight Line. Successively pick pairs of keypoints until the four interior lines shown below are created.

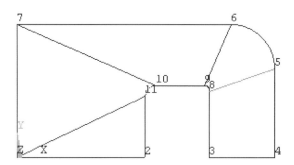

Figure 2-35 Lines added to geometry.

2. Main Menu > Preprocessor > Modeling > Create > Areas > Arbitrary > By Lines
Pick the three lines defining the lower left triangular area. > **Apply** > Repeat for the quadrilateral areas. > **Apply** > **OK**

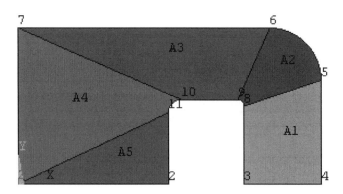

Figure 2-36 Quadrilateral/Triangular regions.

3. Main Menu > Preprocessor > Modeling > Operate > Booleans > Glue > Areas > Pick All

The **glue** operation preserves the boundaries between areas that we will need for mapped meshing.

4. Main Menu > Preprocessor > Meshing > Size Cntrls > ManualSize > Lines > All Lines Enter **4 for NDIV, No. element divisions** > **OK**

All lines will be divided into four segments for mesh creation.

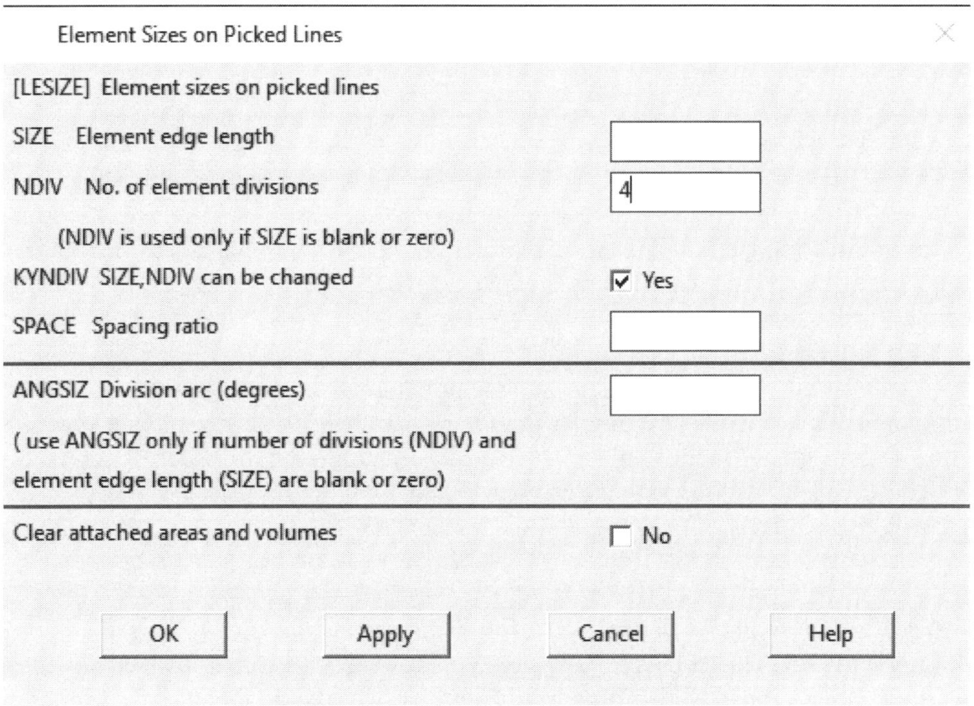

Element Sizes on Picked Lines   ✕

[LESIZE] Element sizes on picked lines

SIZE   Element edge length   [   ]

NDIV   No. of element divisions   [4|]

    (NDIV is used only if SIZE is blank or zero)

KYNDIV SIZE,NDIV can be changed   ☑ Yes

SPACE Spacing ratio   [   ]

ANGSIZ Division arc (degrees)   [   ]

(use ANGSIZ only if number of divisions (NDIV) and
element edge length (SIZE) are blank or zero)

Clear attached areas and volumes   ☐ No

    [OK]   [Apply]   [Cancel]   [Help]

Figure 2-37 Element size on picked lines.

5. Main Menu > Preprocessor > Element Type > Add/Edit/Delete > Add > Solid > Quad 8node   183 > OK (Use the 8-node quadrilateral element for the mesh.)

6. Main Menu > Preprocessor > Meshing > Mesh > Areas > Mapped > 3 or 4 sided > Pick All

The mesh below is created. Applying boundary and load conditions and solving gives the von Mises stress distribution shown. The stress contours are discontinuous because of the poor mesh quality. Notice the long and narrow quads near the point of maximum stress. We need more elements and they need to be better shaped with smaller aspect ratios to obtain satisfactory results.

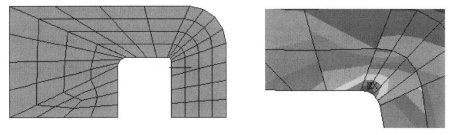

Figure 2-38 Mapped mesh and von Mises results.

One can tailor the mapped mesh by specifying how many elements are to be placed along which lines. This allows much better control over the quality of the mesh, and an example of using this approach is described in Lesson 4.

2-10 CONVERGENCE

The goal of finite element analysis as discussed in this lesson is to arrive at computed estimates of deflection, strain and stress that converge to definite values as the number of elements in the mesh increases, just as a convergent series arrives at a definite value once enough terms are summed.

For elements based on assumed displacement functions that produce continuum models, the computed displacements are smaller in theory than the true displacements because the assumed displacement functions place an artificial constraint on the deformations that can occur. These constraints are relaxed as the element polynomial is increased or as more elements are used. Thus your computed displacements usually converge smoothly from below to fixed values.

Strains are the x and/or y derivatives of the displacements and thus depend on the distribution of the displacements for any given mesh. The strains and stresses may change in an erratic way as the mesh is refined, first smaller than the final computed values, then larger, etc.

Not all elements are developed using the ideas discussed above, and some will give displacements that converge from above. (See Lesson 6.) In any case you should be alert to computed displacement and stress variations as you perform mesh refinement during the solution of a problem.

2-11 TWO-DIMENSIONAL ELEMENT OPTIONS

The analysis options for two-dimensional elements are: **Plane Stress**, **Axisymmetric**, **Plane Strain**, **Plane Stress with Thickness** and **Generalized Plane Strain**. The two examples thus far in this lesson were of the third type, namely problems of plane stress in which we provided the thickness of the part.

The first analysis option, **Plane Stress**, is the ANSYS default and provides an analysis for a part with **unit thickness**. If you are working on a design problem in which the thickness is not yet known, you may wish to use this option and then select the thickness based upon the stress, strain, and deflection distributions found for a unit thickness.

The second option, **Axisymmetric** analysis, is covered in detail in Lesson 3.

Plane Strain occurs in a problem such as a cylindrical roller bearing caged against axial motion and uniformly loaded in a direction normal to the cylindrical surface. Because there is no axial motion, there is no axial strain. Each slice through the cylinder behaves like every other and the problem can be conveniently analyzed with a planar model.

Another plane strain example is that of a long retaining wall, restrained at each end and loaded uniformly by soil pressure on one or more faces.

Generalized Plane Strain allows a nonzero strain the Z direction as well as Z strain variations with X and Y.

2-12 SUMMARY

Problems of stress concentration in plates subject to in-plane loadings were used to illustrate ANSYS analysis of plane stress problems. Free **triangular** and **quadrilateral** element meshes were developed and analyzed. **Mapped meshing** with quads was also presented. Similar methods are used for solving problems involving plane strain; one only has to choose the appropriate option during element selection. The approach is also applicable to axisymmetric geometries as discussed in the next lesson.

2-13 PROBLEMS

In the problems below, use triangular and/or quadrilateral elements as desired. Triangles may produce more regular shaped element meshes with free meshing. The six-node triangles and eight-node quads can approximate curved surface geometries and, when stress gradients are present, give much better results than the four-node quad elements.

2-1 Find the maximum stress in the aluminum plate shown below. Convert the 12 kN concentrated force into an equivalent pressure applied to the edge. Use tabulated stress concentration factors to independently calculate the maximum stress. Compare the two results by determining the percent difference in the two answers.

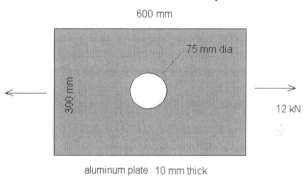

Figure P2-1

2-2 Find the maximum stress for the plate from 2-1 if the hole is located halfway between the centerline and top edge as shown. You will now need to model half of the plate instead of just one quarter and properly restrain vertical rigid body motion. One way to do this is to fix one keypoint along the centerline from UY displacement.

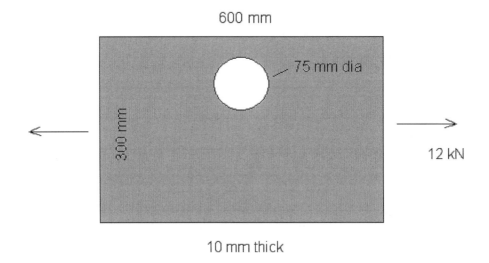

Figure P2-2

2-3 An aluminum square 10 inches on a side has a 5-inch diameter hole at the center. The object is in a state of **plane strain** with an internal pressure of 1500 psi. Determine the magnitude and location of the maximum principal stress, the maximum principal strain, and the maximum von Mises stress. Note that no thickness need be supplied for plane strain analysis.

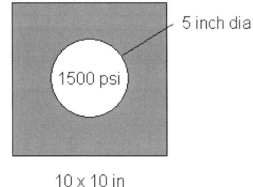

Figure P2-3

2-4 Repeat 2-3 for a steel plate one inch thick in a state of plane stress.

2-5 See if you can reduce the maximum stress for the plate of problem 2-1 by adding holes as shown below. Select a hole size and location that you think will smooth out the 'stress flow' caused by the load transmission through the plate.

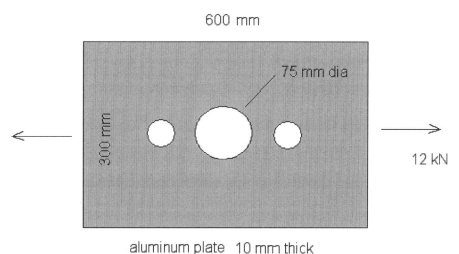

Figure P2-5

2-6 Repeat 2-1 but the object is now a plate with notches or with a step in the geometry. (See the next figure.) Select your own dimensions, materials, and loads. Use published stress concentration factor data to compare to your results. The published results are for plates that are relatively long so that there is a uniform state of axial stress at either end relatively far from notch or hole. Create your geometry accordingly.

Figure P2-6

2-7 Solve the seatbelt component problem of Tutorial 2B again using six node triangular elements instead of the quadrilaterals. Experiment with mesh refinement. Turn on **Smart Sizing** using size controls to examine the effect on the solution. See if you can compute a maximum von Mises stress of around 140 kpsi.

2-8 Determine the stresses and deflections in an object 'at hand' (such as a seatbelt tongue or retaining wall) whose geometry and loading make it suitable for plane stress or plane strain analysis. Do all the necessary modeling of geometry (use a CAD system if you wish), materials and loadings.

2-9 A cantilever beam with a unit width rectangular cross section is loaded with a uniform pressure along its upper surface. Model the beam as a problem in plane stress. Compute the end deflection and the maximum stress at the cantilever support. Compare your results to those you would find using elementary beam theory.

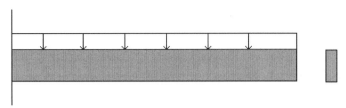

Figure P2-8

Restrain UX along the cantilever support line, but restrain UY at only one keypoint along this line. Otherwise, the strain in the Y direction due to the Poisson effect is prevented here, and the root stresses are different from elementary beam theory because of the singularity created. (Try fixing all node points in UX and UY and see what happens.)

Select your own dimensions, materials, and pressure. Try a beam that's long and slender and one that's short and thick. The effect of shear loading becomes more important in the deflection analysis as the slenderness decreases.

Lesson 3

Axisymmetric Problems

3-1 OVERVIEW

Axisymmetric Analysis – A problem in which the geometry, loadings, boundary conditions and materials are symmetric with respect to an axis is one that can be solved with an axisymmetric finite element model. The tutorial in this lesson discusses

- ♦ Creating model geometry using an IGES file or a text file

- ♦ Evaluating the response of a vessel to an internal pressure

3-2 INTRODUCTION

For problems that are rotationally symmetric about an axis, a slicing plane that contains the symmetry axis exposes the interior configuration of the geometry. Since any slice created by such a plane looks like any other slice, the problem can be conveniently analyzed by considering any one planar section as shown in the figure below.

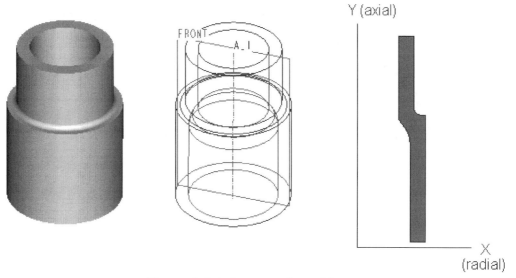

Figure 3-1 Axisymmetric problem.

3-3 CYLINDRICAL PRESSURE VESSEL

A steel pressure vessel with planar ends is subjected to an **internal pressure of 35 MPa** (35 x 10^6 N per m^2). The vessel has an **outer diameter of 200 mm**, an over-all **length of 400 mm** and a **wall thickness of 25 mm**. There is a **25 mm fillet radius** where the interior wall surface joins the end cap as shown in the figure below.

The vessel has a longitudinal axis of rotational symmetry and is also symmetric with respect to a plane passed through it at mid-height. Thus the analyst need consider only the top or bottom half of the vessel.

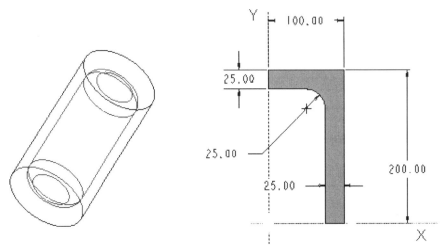

Figure 3-2 Cylindrical pressure vessel.

The global **Y-axis** is the **ANSYS default axis of symmetry for axisymmetric problems**, and **X** is aligned with the radial direction. The **radial stresses** are labeled **SX**, and the **axial stresses** are labeled **SY**. When the vessel is pressurized, it will expand creating stresses in the **circumferential** or **hoop** direction. These are labeled **SZ**.

A ninety-degree slice of the top half of the vessel is created in the **X-Y plane** in a solid modeler for import as an **IGES file** into ANSYS. IGES is a neutral CAD format file definition. Each software manufacturer writes software for translating from IGES to its native file format and vice versa. The angular extent of the slice is not important; we just want to expose the X-Y plane for analysis. (See the instructions below if you want to create the geometry using ANSYS commands instead of using an IGES file from a solid modeler.)

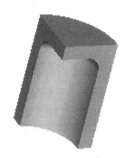

Figure 3-3 Solid model.

The material properties for the vessel are elastic modulus E = 200 GPa (2 x 10^{11} N per m^2), Poisson's ratio = 0.3, Yield strength = 330 MPa (330 x 10^6 N per m^2). The geometric modeling was performed using mm as units of length, so we will use a consistent set of units with **E = 2 x 10⁵ N per mm²** and **Yield strength = 330 N per mm²**. An internal **pressure of 35 N per mm²** will be used. Results calculated with these inputs will have displacements in mm and stresses in N per mm² (MPa).

3-4 TUTORIAL 3A – PRESSURE VESSEL

Objective: Determine if yielding will occur in the cylindrical pressure vessel described above if it is pressurized as indicated.

PREPROCESSING

1. Start ANSYS; set the working directory and job name as usual. First develop the geometry for the vessel cross section.

If you are not using an IGES file to define the geometry for this exercise, you can create the **2D cross section** geometry directly in ANSYS with key points, lines, and arcs by selecting File > Read Input from and reading the text file below.

```
/FILNAM,Tutorial3A
/title, Pressure Vessel Geometry
/prep7
! Create geometry

k, 1,  75.,   0.          ! Keypoint 1 is at 75.0, 0.0
k, 2, 100.,   0.
k, 3, 100., 200.
k, 4,   0., 200.
k, 5,   0., 175.
k, 6,  50., 175.
k, 7,  75., 150.
k, 8,  50., 150.

L, 7, 1                   ! Line from keypoints 7 to 1
L, 1, 2
L, 2, 3
L, 3, 4
L, 4, 5
L, 5, 6

larc, 7, 6, 8, 25.        ! Fillet arc

AL, 1, 2, 3, 4, 5, 6, 7   ! Area defined by lines 1 thru 7
```

Skip to the element selection phase, step 4 below.

Figure 3-4 Cross section from script.

Alternatively, import the IGES file of the vessel slice. The **results shown** in this tutorial were created using an imported **IGES file** from a solid modeler.

2. Utility Menu > File > Import > IGES Select the IGES file you created earlier. Accept the ANSYS import default settings. If you have trouble with the import, select the alternate options and try again. Defeaturing is an automatic process to remove inconsistencies that may exist in the IGES file such as lines that, because of modeling or the translation process, do not quite join.

Turn the solid model around to show the X-Y plane (Ctrl-Right mouse)

3. Utility Menu > PlotCtrls > Pan, Zoom, Rotate > Back

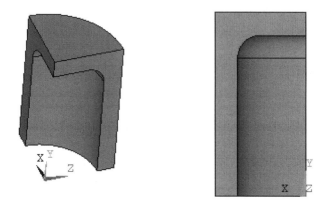

Figure 3-5 X-Y plane of cross section.

Select the **six-node triangular element** for this analysis and choose the **axisymmetric** modeling option.

4. Main Menu > Preprocessor > Element Type > Add/Edit/Delete >Add > Structural Solid > Quad 8node 183 > OK

5. Options (Element shape K1) > **Triangle,**

Options (Element behavior K3) > **Axisymmetric > OK > Close**

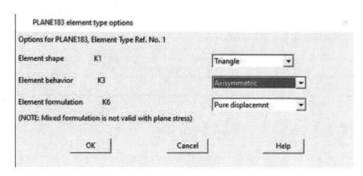

Enter the material properties

6. Main Menu > Preprocessor > Material Props > Material Models

Material Model Number 1, click **Structural > Linear >Elastic > Isotropic**

Enter **EX = 2.0E5** and **PRXY = 0.3 > OK**

Now mesh the X-Y plane area. (Turn area numbers on; **Plot** > **Areas**; then to select an area, click its number. Use **Cntl** + **Mouse** buttons to dynamically manipulate the image.)

7. Main Menu > **Preprocessor** > **Meshing** > **Mesh** > **Areas** > **Free** Pick the cross section area in the X-Y plane > **OK** (Skip to Step 10 if you are not using an IGES file.)

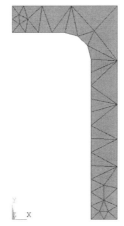

Figure 3-6 Initial mesh.

The IGES solid model is no longer needed and should be deleted so its lines and areas will not interfere with subsequent operations.

8. Main Menu > **Preprocessor** > **Modeling** > **Delete** > **Volume and Below**   (Don't be surprised if everything disappears. Just **Plot** > **Elements** to see the mesh again.)

9. Utility Menu > **PlotCtrls** > **Pan, Zoom, Rotate** > **Front**   (To see the front side of the mesh.)

Increase the mesh density to obtain good accuracy in the solution.

10. Main Menu > **Preprocessor** > **Meshing** > **Modify Mesh** > **Refine At** > **All** > **Level of Refinement   2** > **OK**

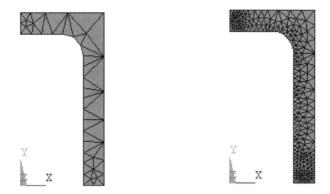

Figure 3-7 Initial and Refined mesh.

Apply displacement and pressure boundary conditions. Select the bottom horizontal edge and restrain it from UY movement.

11. Main Menu > Preprocessor > Loads > Define Loads > Apply > Structural > Displacement > On Lines Pick the lower edge > **UY = 0.** > **OK**

No displacement constraint is required on the top left vertical edge since it is on the axis of symmetry and axisymmetric modeling prevents any radial motion here.

Apply a pressure to the inside of the cylinder.

12. Main Menu > Preprocessor > Loads > Define Loads > Apply > Structural > Pressure > On Lines

Select the three interior lines and set **pressure = 35** > **OK**

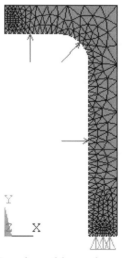

Figure 3-8 Loads and boundary conditions.

(The pressure may show as single or multiple arrows depending upon the solution stage. If you lose it completely, it will resurface during solution or do a **Utility Menu > Loads > List**.)

SOLUTION

13. Main Menu > Solution > Solve > Current LS > OK

POSTPROCESSING

We can now **plot the results** of this analysis and **list the computed values**.

14. Main Menu > General Postproc > Read Results > First Set

Plot Results > Contour Plot > Element Solu > Stress > von Mises stress > OK

A maximum von Mises stress of just over 200 N per mm^2 (200MPa) is computed at the fillet where the interior sidewall and top join. (Your meshes and results may differ a bit from those shown.)

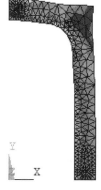

Figure 3-9 Von Mises Stress on deformed shape.

The deformed shape looks correct, so let's look more closely at the computed solution.

Theoretical Solution

The stresses at **points removed from the stress concentration** at the end caps (say at the lower portion of the segment shown in the figure above) could be predicted using **thick walled cylinder** theory. For the geometry and loading of this example, **theory** predicts the following stress components at the **inner surface** of the cylinder.

Radial Stress (SX) = -35 N per mm^2

Axial Stress (SY) = 45 N per mm^2

Hoop Stress (SZ) = 125 N per mm^2

Radial Deflection = 0.0458 mm

These theoretical results can be used to evaluate the ANSYS solution. Zoom in on the **bottom portion** of the model. Also turn the node numbering on and plot the elements.

The **element solution** plot of the von Mises stress shows smooth contours indicating a reasonably accurate solution, so we can safely use **nodal solution stress values**. List the values for node 20. (The node numbers in your model will depend upon the model building and meshing sequence you use and may be different from what is shown here.)

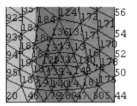

Figure 3-10 Model detail.

15. Main Menu > General Postproc > List Results > Nodal Solution > Stress > X-Component of stress > OK

```
PRINT S    NODAL SOLUTION PER NODE

 ***** POST1 NODAL STRESS LISTING *****
 PowerGraphics Is Currently Enabled

 LOAD STEP=     1  SUBSTEP=      1
  TIME=    1.0000     LOAD CASE=   0
 NODAL RESULTS ARE FOR MATERIAL    1

 THE FOLLOWING X,Y,Z VALUES ARE IN GLOBAL COORDINATES

   NODE    SX         SY          SZ          SXY
. . . . . .
     20  -34.846    46.440      127.37      -0.18278E-01
. . . . . .
```

The computed values for SX, SY and SZ are only a few percent different from the theoretical solution. (Your numerical values may be slightly different from those shown above.) This correlation gives us confidence in the solution, and if we refine the mesh in this area, closer agreement in the radial (SX) and hoop (SZ) stresses could be obtained. The theoretical solution for the axial stress (SY) is an average value (P/A), so some disagreement is to be expected, and the finite element solution probably gives a better picture of the true stress distribution. (The shear stresses SYZ and SXZ are zero for axisymmetric problems and were deleted from the listing above to save space.) We also check the radial displacement result.

16. Main Menu > General Postproc > List Results > Nodal Solu > DOF Solution > X-Component of displacement > OK

```
 PRINT U    NODAL SOLUTION PER NODE

 ***** POST1 NODAL DEGREE OF FREEDOM LISTING *****

 LOAD STEP=     1  SUBSTEP=      1
  TIME=    1.0000      LOAD CASE=   0
```

```
   THE FOLLOWING DEGREE OF FREEDOM RESULTS ARE IN GLOBAL
COORDINATES

    NODE       UX
. . . .
     20  0.46524E-01
. . . .
```

The computed radial displacement at node 20 is only 0.77 percent different from the theoretical.

To evaluate the possibility of **yielding** for this ductile material we compare the **von Mises** stress to the material yield strength. The maximum vessel stresses occur at the fillet where the vertical wall and the end caps join. Zoom in on an **element solution plot** of **SEQV** in the vicinity of the fillet, and we see that the stress contours are not completely smooth; there are discontinuities between elements. To get better accuracy, remove the pressure and refine the mesh using the fillet geometry to indicate the region to be refined.

17. Main Menu > Preprocessor > Loads > Define Loads > Delete > Structural > Pressure > On Lines   Main Menu > Preprocessor > Meshing > Modify Mesh > Refine At > Lines (Select the fillet.) **Level = 1 > OK** (Removing the loads prior to remesh is not always necessary.)

(To get a clearer view of the node numbers, you can turn off the MX, MN symbols **PlotCtrls > Window Controls > Window Options > MINM Min-Max Symbols > Off > OK**)

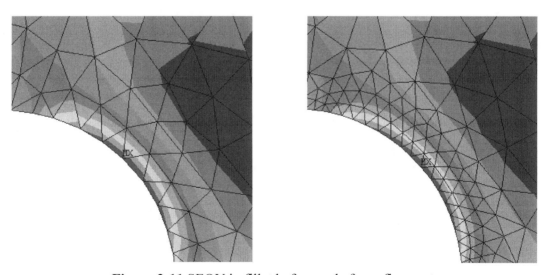

Figure 3-11 SEQV in fillet before and after refinement.

The maximum von Mises stress occurs at the location shown with a value of about **205 N per mm^2**. Comparing this with the yield strength of 330 N per mm^2, we conclude that yielding is not a problem at the fillet for this loading.

Use the **Results Viewer** to inquire further at the point of maximum von Mises stress.

18. Main Menu > Postproc > Results Viewer.

19. Choose a result item > Element Solution > Stress > von Mises.

20. Select **Contour**.

21. Click on the first icon, **Plot Results,** ; the figure below is displayed.

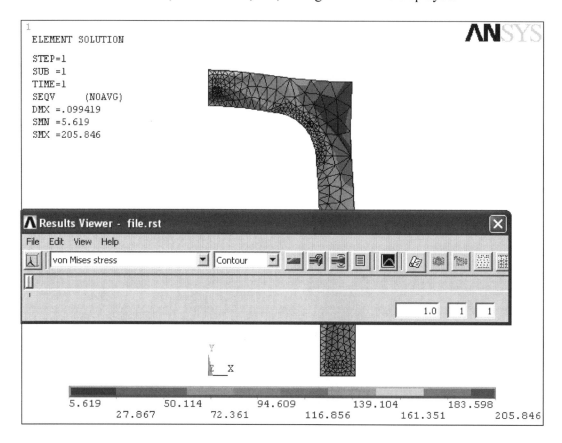

Figure 3-12 Results viewer.

22. Zoom in on the region of maximum stress, then click on the **second icon, Query Results** and click on the node where the maximum value is indicated. The figure to the right is shown. The query value is about 204 MPa.

'**Multiple items at this location**' may be shown in a multiple entities window because there are three elements that join at this node, and different stress levels may be calculated in each element. For a continuum problem such as the one we are analyzing here, the magnitude of difference in element solution quantities is an indication of the potential error in the solution.

23. OK > Close

Now from the pull-down list in the **Results Viewer** select **Nodal Solution** (instead of Element Solution) > **Stress** > **von Mises**. **Plot Results**, then **Query Results** again at the point of maximum stress.

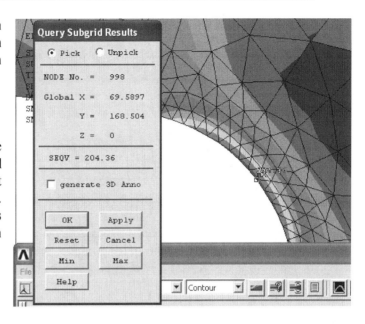

Figure 3-13 Query results, element solution.

(To return to the ANSYS main menu, close the Results Viewer window when you are finished.)

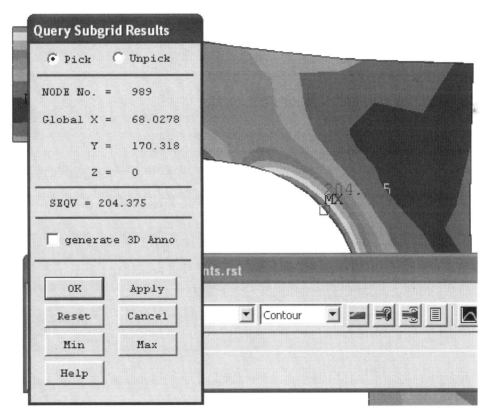

Figure 3-14 Query results, nodal solution

The nodal solution magnitudes are the average values computed using quantities calculated for each element connected to the node. Thus the nodal solution contours are always nice and smooth. They give a good representation of stress variations once the mesh is sufficiently dense to produce small differences in quantities calculated for elements with common nodes. (Your results might differ slightly from those shown here.)

3-5 OTHER ELEMENTS

We used triangular elements for meshing in the examples above, but the axisymmetric option is also available with the other two-dimensional ANSYS elements: the four-node quadrilateral (42 or 182) and the eight-node quadrilateral (82 or 183). Quadrilateral elements sometimes offer advantages over triangles when developing mapped meshes or when rectangular or trapezoidal areas can be identified for free meshing. For arbitrary shapes, free triangular meshes may produce more regular shaped elements than free quad meshes. **Smart Size** controls also can be used very effectively to control mesh quality (see Lesson 4).

3-6 SUMMARY

A pressure vessel was used as an example of a problem with symmetry in geometry, boundary conditions, loadings, and material properties about a central axis. An upset shaft in tension is one of many other cases in which simple two-dimensional models may be used to great advantage in the solution of important engineering problems possessing symmetry. The problems below further explore this type of modeling process.

3-7 PROBLEMS

3-1 The steel pressure vessel shown has an internal pressure of 4000 psi. The overall length is 50 inches with inside and outside radii of 5 and 10 inches respectively. Compare the computed stress components at the mid section to theoretical values. Also determine if the vessel will yield with the given internal pressure if its yield strength is 50 kpsi.

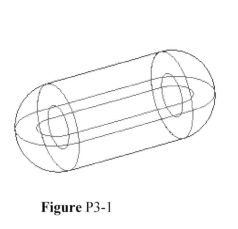

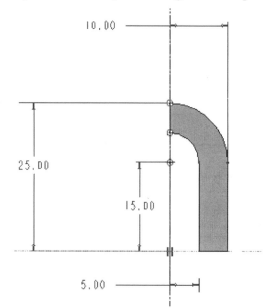

Figure P3-1

3-2 Determine the stresses and deflections of the vessel geometry of problem 3-1 if the dimensions are measured in cm and it is subjected to an **external pressure** corresponding to submersion in 100 m of water.

3-3 Compute the maximum tensile stress in a circular shaft with an upset shoulder under **axial loading**. Compare your results to those one can compute using tabulated stress concentration factors for this geometry. Select your own diameters, fillet radius, length, materials, and loads.

Figure P3-3

3-4 The anchor device below is loaded by an axial force of 10,000 lbs. Dimensions shown are in inches. Apply the force as a pressure to the top surface. The base is buried so that its upper surface cannot move in the Y direction and its vertical edge surface cannot move radially.

Select a gray cast iron from which to make the part and find its material properties. Use the maximum normal stress theory of failure to determine if failure is likely to occur in this part. That is, compare the maximum computed principal tensile stress, **S1**, with the material ultimate strength. Use **General Postprocessor > Plot Results > Vector Plots > Predefined > Stress Principal** and sketch the fracture plane if failure were to occur. The groove is 0.20 inch in **diameter**.

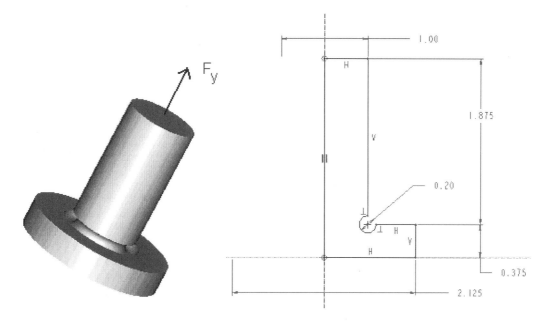

Figure P3-4

3-5 A circular aluminum plate 1 m in diameter and 5 cm in thickness is loaded by a uniform pressure of 0.5 Pa distributed over a 0.25 m diameter circle at the center of the plate. The outer edges of the plate are completely fixed from motion. Find the maximum deflection and von Mises stress and their locations.

3-6 Solve 3-5 again if the plate thickness varies from 4 cm at the outer edge to a constant 7 cm over the 0.25 m central circle.

3-7 Compute the magnitude of the maximum radial deflection and the maximum stresses in a hollow sphere with internal/and or external pressure. Select your own dimensions, materials and pressures. Compare your results with published theoretical values. You can create a quadrant of the section using the ANSYS commands **Preprocessor > Create > Circle > Partial Annulus**.

Lesson 4

3-Dimensional Problems

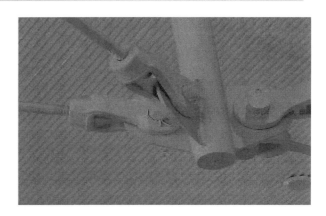

4-1 OVERVIEW

Three-Dimensional Analysis – Problems that do not have an appropriate structural, plane stress, plane strain or axisymmetric model must be solved as three-dimensional problems. Lesson 4 discusses:

- ◆ Developing 3-D models from solid modelers.

- ◆ Meshing volumes with tetrahedron or brick elements.

- ◆Using cyclic symmetry modeling when applicable.

4-2 INTRODUCTION

Three-dimensional regions that define mechanical parts are often solved using solid tetrahedron or brick elements when simpler models are not available or sometimes when the development of a simpler model involves an excessive commitment of engineering time.

In this lesson we will revisit the pressure vessel of Lesson 3, treating it as a three-dimensional problem instead of an axisymmetric problem. We create the ANSYS model geometry by importing the IGES file of the pressure vessel that was created in a solid modeler. We also give instructions for creating this volume directly in ANSYS in case you are not using a solid modeler in conjunction with this exercise.

4-3 CYLINDRICAL PRESSURE VESSEL

Figure 4-1 shows the cylindrical pressure vessel of Lesson 3. Even though we are going to use three-dimensional modeling for this analysis, we can still take advantage of symmetry to reduce the size of the model that must be analyzed.

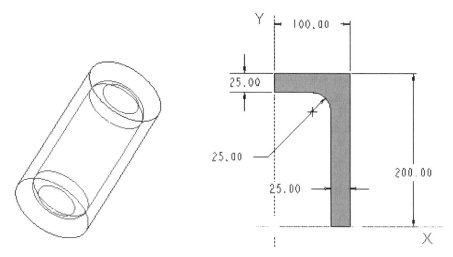

Figure 4-1 Cylindrical pressure vessel.

We will use the 90-degree segment of the solid model of the vessel as we did in Lesson 3. The symmetric nature of the geometry and loading means that displacements are zero in directions normal to the faces exposed by the vertical and horizontal cuts employed to create this one-eighth segment of the cylinder.

Figure 4-2 Solid.

4-4 TUTORIAL 4A – PRESSURE VESSEL

Objective: Analyze the pressure vessel as a three dimensional solid using ten-node tetrahedron elements. Compare computed results with thick walled cylinder theory and the maximum von Mises stress with that calculated using the axisymmetric model of Lesson 3.

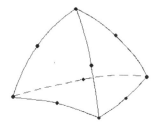

Figure 4-3 10-node tet.

PREPROCESSING

1. Start ANSYS; set the working directory and job name as usual.

2. If you are not using an IGES file to define the geometry for this exercise, you can create the geometry directly in ANSYS with key points, lines, and arcs by selecting **File > Read Input from** and reading the text file below (then skip step 3).

```
/FILNAM,Tutorial4A
/title, Pressure Vessel Geometry
/prep7
! Create geometry

k, 1,  75.,   0.        ! Keypoint 1 is at 75.0, 0.0
k, 2, 100.,   0.
k, 3, 100., 200.
k, 4,   0., 200.
k, 5,   0., 175.
k, 6,  50., 175.
k, 7,  75., 150.
k, 8,  50., 150.

L, 7, 1                 ! Line from keypoints 7 to 1
L, 1, 2
L, 2, 3
L, 3, 4
L, 4, 5
L, 5, 6

larc, 7, 6, 8, 25.      ! Fillet arc

AL, 1, 2, 3, 4, 5, 6, 7  ! Area defined by lines 1 thru 7
! rotate the area about k4-k5 axis through 90 deg.

vrotat, all, , , , , , 4, 5, 90.
```

Alternatively, import the IGES file for the 90-degree segment of the pressure vessel.

3. Utility Menu > File > Import > IGES Select the IGES file you created earlier. Accept the ANSYS import default settings. If you have trouble with the import, select the alternate options and try again. **Defeaturing** is an automatic process to remove inconsistencies that may exist in the IGES file because of modeling or the IGES translation process. (**Ctrl-Right** mouse button to spin the volume.)

Select the **10-node tetrahedron** element (number 187 in the ANSYS menu) for this analysis. The 10-node tets have mid-side nodes to facilitate modeling curved surfaces.

4. Main Menu > Preprocessor > Element Type > Add/Edit/Delete > Add > Solid > Tet 10 node   187 > OK

Enter the material properties

5. Preprocessor > Material Props > Material Models

Material Model Number 1, Double click **Structural > Linear > Elastic > Isotropic**

Enter **EX = 2.0E5** and **PRXY = 0.3 > OK**

Figure 4-4 Solid in ANSYS.

Now use the default settings to mesh the volume that has been created or imported.
6. Main Menu > Preprocessor > Meshing > Mesh > Volumes > Free Pick the solid. **> OK**

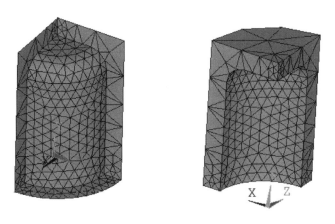

Figure 4-5 Initial tetrahedron mesh.

Now apply the displacement boundary conditions. We will apply these to the boundary faces of the solid to facilitate mesh modification operations. Displacements perpendicular to the left, right, and bottom faces exposed by the planes of symmetry must be zero. Use **PlotCtrls > Numbering > Node numbers > On > OK**. Plot Areas and pick the Area No.

7. Main Menu > Preprocessor > Loads > Define Loads > Apply > Structural > Displacement > On Areas Pick the exposed face on the right. **> OK > UX = 0. > OK**

8. Main Menu > Preprocessor > Loads > Define Loads > Apply > Structural > Displacement > On Areas Pick the exposed face on the left. **> OK > UZ = 0. > OK**

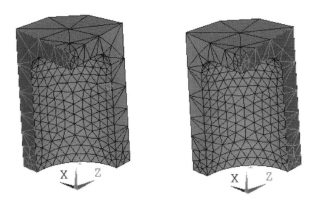

Figure 4-6 The UX = 0 and UZ = 0 faces.

Set UY = 0 on the bottom face. Use **control** + **right mouse** to dynamically rotate object so you can see the bottom face.

9. Main Menu > Preprocessor > Loads > Define Loads > Apply > Structural > Displacement > On Areas Pick the bottom face. **> OK > UY = 0. > OK**

Apply the pressure to the inside of the cylinder

10. Main Menu > Preprocessor > Loads > Define Loads > Apply > Structural > Pressure > On Areas Select the three areas that make up the inside surface. **> OK** Set **pressure = 35 > OK**

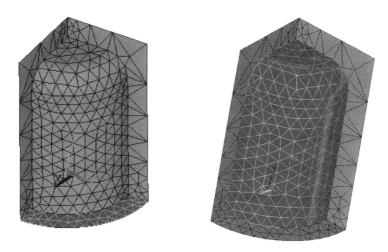

Figure 4-7 The UY = 0 and applied Pressure faces.

Check your work by listing the loads. PlotCtrls > Numbering > AREA Area numbers On

11. Utility Menu > List > Loads > DOF Constraints > On All Areas

12. Utility Menu > List > Loads > Surface > On All Areas

You can also check the model size while in the Preprocessor

SOLUTION

13. Main Menu > Solution > Solve > Current LS > OK

POSTPROCESSING

First plot the von Mises **element stress solution** to get an idea of the accuracy of the initial mesh. Your meshes/results may differ slightly from those shown.

14. Main Menu > General Postproc > Read Results > First Set

15. Main Menu > General Postproc > Plot Results > Contour Plot > Element Solu > Stress > von Mises stress > OK

The element solution plot shows discontinuous stress contours across element boundaries indicating poor accuracy for this coarse mesh. The maximum computed value is around 224 N per mm^2 (222 MPa), which is about 9 percent higher than the value calculated with the axisymmetric model in Lesson 3. Also we note that because of the symmetry of the problem, the contours should be smooth in the circumferential direction.

Figure 4-8 Von Mises stress.

Clearly we will need a finer mesh in order to have confidence in the results. Ordinarily, of course, the results are not known in advance as in this case, and we must employ previous experience and good judgment to arrive at satisfactory results.

Return to the Preprocessor and refine the mesh. It may be necessary to remove the pressures from areas before remeshing. It is **important** to **save your work before remeshing**. The remesh could request more nodes and elements than are available with your license, and ANSYS may **shut down**. For other options, see **4-5 Smart Sizing** below.

16. Main Menu > Preprocessor > Loads > Define Loads > Delete > Structural > Pressure > On Areas Select the three areas that make up the inside surface. **> OK > OK.** (Deleting the loads before remeshing is not always necessary. Check your installation.)

17. Main Menu > Preprocessor > Meshing > Modify Mesh > Refine at > All > Level of Refinement 1 > OK

After refining the mesh everywhere, refine the area at the fillet on the inside of the cylinder.

18. Main Menu > Preprocessor > Meshing > Modify Mesh > Refine At > Areas (pick the fillet area) **> Level of Refinement 1 > OK**

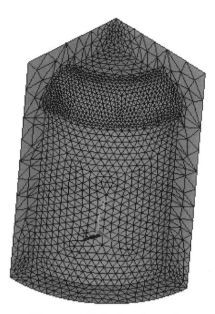

Figure 4-9 Refined mesh.

The remesh operation produces a model with about **10,000 elements** and **16,000 nodes**.

Reapply the loads if you deleted them.

19. Main Menu > Preprocessor > Loads > Define Loads > Apply > Structural > Pressure > On Areas

Select the three areas that make up the inside surface and set **pressure = 35 > OK**

Return to the SOLUTION. The solution may now take a perceptible amount of time depending upon the speed of your computer. Use the Postprocessor to plot the von Mises stress and zoom in on the fillet region.

Using the **Results Viewer**, the maximum von Mises stress is found to be about 210 N per mm^2 with this mesh, and while the element solution stress contour plots are not as smooth as one might like, these results are fairly accurate. If we continue with the mesh refinement, we would expect to see the stress levels converge to some stable values.

It is also important to check the stresses at points away from the fillet to see how they compare with theoretical values.

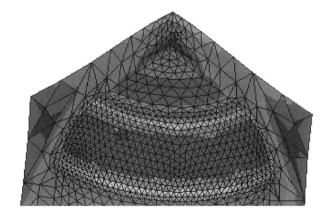

Figure 4-10 Von Mises stress.

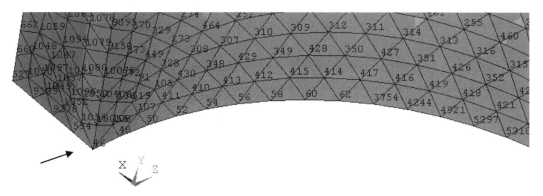

Figure 4-11 Node numbering at inner surface.

The node at the inside surface on the left is number 46 (your numbering scheme may be different). In this location SX corresponds to the radial stress component, SY the axial component, and SZ the hoop component.

Theoretical Solution:

$$\text{Radial Stress (SX)} = -35 \text{ N per mm}^2$$

$$\text{Axial Stress (SY)} = 45 \text{ N per mm}^2$$

$$\text{Hoop Stress (SZ)} = 125 \text{ N per mm}^2$$

ANSYS Solution: (List SX)

```
   NODE     SX          SY       SZ       SXY           SYZ            SXZ

   46   -34.913    46.496    127.51   -0.21380E-001  -0.64364E-002  -0.18287
```

And the computed results are seen to have good accuracy.

4-5 SMART SIZE

There are a number of parameters and options available in ANSYS for use during mesh creation. One of these is the **smart size** option that employs meshing rules to adjust the element density to the geometry of the model. Either clear the mesh from the model used in the previous tutorial (**Preprocessor > Meshing > Clear > Volumes**) or start the problem again and mesh the volume again. In either case, turn smart size on before creating the initial mesh.

1. Main Menu > Preprocessor > Meshing > Size Cntls > Smart Size > Basic > LVL Size Level > 3 > OK

2. Main Menu > Preprocessor > Meshing > Mesh > Volumes > Free > (pick the solid) **> OK**

This operation creates the mesh with around 700 elements and 1400 nodes shown on the left in the figure below. This mesh has a much better element shape and distribution through the thickness of the vessel than the mesh of Figure 4-5 that was created using the default setting with smart size off.

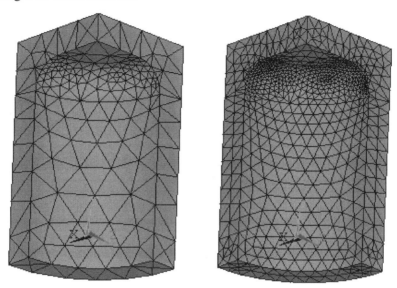

Figure 4-12 Smart size meshes.

The mesh on the right was obtained from **Modify Mesh > All > LEVEL Level of refinement 1**. It has around 5400 elements and 9000 nodes and produces a maximum von Mises stress comparable to that found earlier.

4-6 CYCLIC SYMMETRY

The 3-D pressure vessel problem can be solved with a much smaller model than those above. With the 90-degree slice, the boundary restraints on exposed faces are conveniently aligned with global coordinate directions X, Y, Z, but we can use a much smaller slice if the displacements normal to the exposed face can be set to zero.

We illustrate the ideas by solving the problem again working with a **15-degree** slice defined in **global cylindrical coordinates**.

4-7 TUTORIAL 4B – USING CYCLIC SYMMETRY

1. Start ANSYS and establish jobname and working directory as usual.

2. Import a 15-degree solid model of the pressure vessel or use ANSYS commands to create the 15-degree wedge. Select the **ten-node tetrahedron**, element number **187**. Use the same material properties as before, **EX = 2.0E5** and **PRXY = 0.3**.

3. Utility Menu > WorkPlane > Change Active CS (Coordinate System) to > Global Cylindrical Y

The angular coordinate theta corresponds to a rotation about the Y-axis.

Displacement Boundary conditions are applied to restrain displacement in the **Z direction** on the **X-Y plane (left face)** of the wedge.

4. Main Menu > Preprocessor > Loads > Define Loads > Apply > Structural > Displ > On Areas Select the X-Y face (left side in figure below) and set **UZ = 0.**

On the symmetric face of the wedge apply Symmetric Displacement conditions.

5. Main Menu > Preprocessor > Loads > Define Loads > Apply > Structural > Displacement > Symmetry B.C. > On Areas (Select the symmetric face, right side in figure) > **OK**

The S's in the figure below indicate the symmetric boundary conditions.

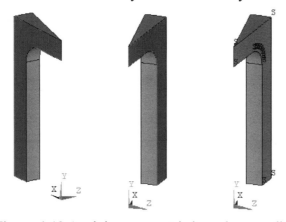

Figure 4-13 Applying symmetric boundary conditions.

Restrain the lower surface of the wedge from displacement in the **Y direction (UY = 0)**, and then apply a **pressure** of 35 N per mm^2 to the interior surface. Create the mesh.

6. Main Menu > Preprocessor > Meshing > Size Cntls > Smart Size > Basic > LVL Size Level > 6 (default) > **OK**

7. Main Menu > Preprocessor > Meshing > Mesh > Volumes > Free > (pick the solid) **> OK**

8. Modify Mesh > All > LEVEL Level of refinement 1 > OK

9. Modify Mesh > Areas > (pick fillet area) **LEVEL Level of refinement 1 > OK**

10. Now **solve** and **plot** the element solution von Mises stress contours. Use Results Viewer. The maximum von Mises stress is about 208 N per mm², about the same as our best estimate from the axisymmetric model. This mesh contains about 1300 elements and 2500 nodes, around 15 percent of the size of the 90-degree model and displays better accuracy.

Figure 4-14 Wedge Von Mises stresses.   **Figure 4-15** Flange.

The approach just described is particularly useful for the analysis of items with **cyclic symmetry** such as the flange shown in the figure above. If the geometry, boundary conditions, loadings, and materials are symmetric with respect to an axis of revolution, we can use a **sector**, say from the center of one hole to the center of the next hole, for analysis and use symmetric boundary conditions when building the model. (Alternatively, you could use a sector centered about a hole.)

4-8 BRICKS

The ANSYS element library also includes **eight-node (SOLID185)** and **twenty-node (SOLID186) hexahedra** (bricks) for three-dimensional modeling. These elements can collapse into wedge, pyramid, or tetrahedron shaped elements. The straight-sided eight-node elements can be used for simple geometries with straight edges.

Figure 4-16 20-node brick elements.

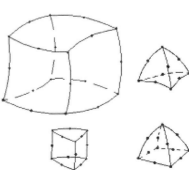

However, the twenty-node element permits modeling of curved surfaces and provides much greater accuracy than the eight-node element. The twenty-node element is shown in the accompanying figure. It is easy to visualize filling a six-faced region with brick elements, but a completely arbitrarily shaped solid can sometimes be difficult to mesh this way.

Mapped Meshes: ANSYS provides a number of meshing methods. With the mapped method, a regular pattern of bricks is mapped onto a six, five or four-faced solid. Examples are shown below.

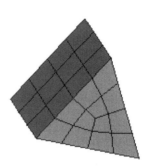
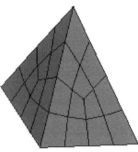

Figure 4-17 Mapped mesh examples.

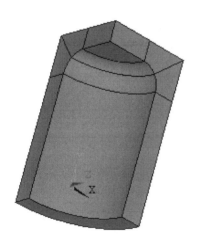

To utilize mapped meshing techniques, the solid must have identifiable four, five or six-sided sub-volumes. If these do not exist, they may be created by concatenating areas, adding planes, etc. Since they have fewer elements and nodes, mapped brick meshes solve faster than equivalent tetrahedron meshes, but their creation can often require more engineering time.

We return to the pressure vessel to illustrate the mapped meshing process. The basic 90-degree solid has been divided into **three six-sided** and **one five-sided** component solids and those components **glued** together to preserve the boundary planes between them.
In the next tutorial we mesh and solve this problem using twenty-node brick elements.

Figure 4-18 Vessel solid.

4-9 TUTORIAL 4C –MAPPED MESH PRESSURE VESSEL

The geometry is developed by using keypoints and lines to create areas and by sweeping the areas through 90 degrees to form the volumes. The volumes are then glued together.

1. Start ANSYS, etc.

2. Read the following text file of keypoints and lines using **File > Read Input from . . .**

```
/FILNAM,Tutorial4C
/title, Pressure Vessel Geometry Creation for Mapped Mesh

/prep7
! Create geometry
k, 1,  75.,   0.          ! Keypoint 1 is at 75.0, 0.0
k, 2, 100.,   0.
k, 3, 100., 150.
k, 4, 100., 200.
k, 5,  50., 200.
k, 6,   0., 200.
k, 7,   0., 175.
k, 8,  50., 175.
k, 9,  75., 150.
k, 10,  50., 150.

x11 = 50. + 25.*cos(45/57.29578)    ! Coords of k11
y11 = 150. + 25.*sin(45/57.29578)

k, 11, x11, y11

L, 9, 1                   ! Line from keypoints 9 to 1
L, 1, 2
L, 2, 3
L, 3, 4
L, 4, 5
L, 5, 6
L, 6, 7
L, 7, 8
L, 9, 3
L, 8, 5
L, 4, 11

larc, 9, 11, 10, 25.   ! Fillet arc
larc, 11, 8, 10, 25.

al, 1, 2, 3, 9           ! Area defined by lines 1,2,3,9
al, 9, 4, 11, 12
al, 13, 11, 5, 10
al, 8, 10, 6, 7

! rotate all areas about k6-k7 axis through 90 deg.

vrotat, all, , , , , , 6, 7, 90.
! glue all volumes together

vglue,all
```

For more information on these commands, search the ANSYS help file. The geometry created by this text file is shown below. (Alternatively use your CAD system to manipulate the geometry.)

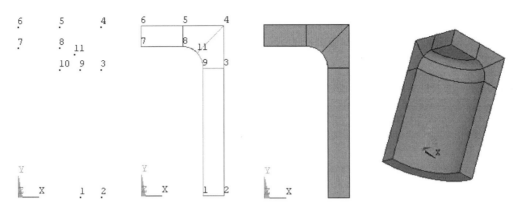

Figure 4-19 Pressure vessel keypoints, lines, areas, and volumes.

Select the twenty-node brick element.

3. Main Menu > Preprocessor > Element Type > Add/Edit/Delete > Add > Solid > Brick 20 node186 > OK

4. Enter material properties **EX = 2.0E5** and **PRXY = 0.3** as before.

Mesh each of the four volumes separately. Use size controls to specify the number of elements along each edge: 2 through the wall thickness, 10 over the 0 to 90 degree arc, and 4 along the remaining edge. Start with the large lower volume.

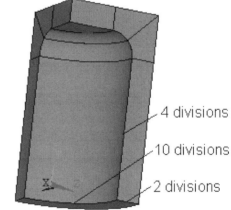

Figure 4-20 Edge divisions.

5. Main Menu > Preprocessor > Meshing > Size Cntrls > Manual Size > Lines > Picked Lines > (pick the wall thickness dimension) > OK Enter NDIV No. element divisions = 2 > OK (KYNDIV SIZE,NDIV can be changed to **no** if need be.)

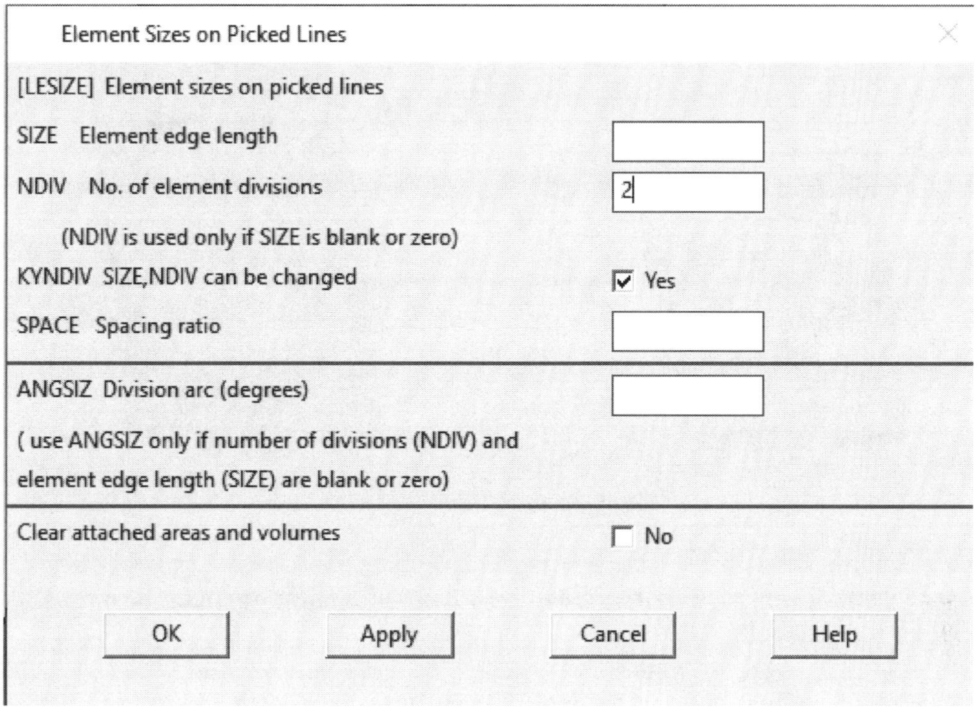

Figure 4-21 Element sizes on picked lines.

6. Repeat step 5 for the other two edges giving them 4 and 10 divisions each.

Then complete the mapping for this component.

7. Main Menu > Preprocessor > Meshing > Mesh > Volumes > Mapped > 4 to 6 sided. Pick the lower solid component. The mesh shown on the left below is created.

8. Repeat step 5 for the two circular arc edges of the fillet.

Only a minimum number of edge division definitions need to be made; compatibility along glued boundary edges will be enforced during the meshing operation.

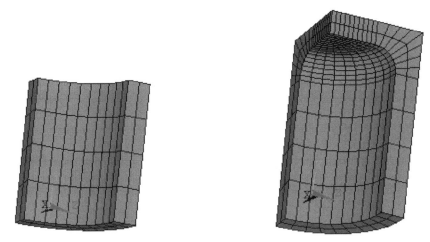

Figure 4-22 Mapped meshes.

9. Repeat the meshing process for the other three solid components. The mesh on the above right is developed.

10. Apply displacement boundary conditions and pressures as before and Solve.

This model has about 390 elements and 2366 nodes and is solved in a few seconds. The maximum von Mises stress is around 11 percent lower than our best estimate from the axisymmetric solution, so we need a higher density mesh in several regions to obtain better accuracy.

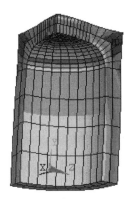

Figure 4-23 Von Mises stress contours.

4-10 SWEEPING

Three-dimensional meshes can also be created for volumes by sweeping a two-dimensional mesh along or about a line. One must select a 2-D element for the area and a 3-D element for the volume. The 2-D mesh is temporary and can be automatically eliminated after the solid mesh is created. To illustrate, consider the pressure vessel one last time.

4-11 TUTORIAL 4D – SWEPT MESH

1. Create the volumes as in Tutorial 4C.

Select the 8-node quad element to mesh the area and the 20-node brick for the volume.

2. Main Menu > Preprocessor > Element Type > Add/Edit/Delete > Add > Solid > Quad 8 node 183 > OK

3. Main Menu > Preprocessor > Element Type > Add/Edit/Delete > Add > Solid > Brick 20 node186 > OK

We will create a mapped 2-D mesh on the lower part of the solid. Use size controls to specify **three** elements through the thickness and **six** elements along the height direction.

4. Main Menu > Preprocessor > Meshing > Size Cntrls > Manual Size > Lines > Picked Lines > (pick the wall thickness dimension *in the X-Y plane*.) **> OK. Enter 3 for NDIV No. element divisions > OK**

5. Do the same for the height with NDIV = 6

6. Main Menu > Preprocessor > Meshing > Mesh > Areas > Mapped > 3 to 4 sided
Pick the area on the left vertical edge of the solid. The quad mesh below is created.

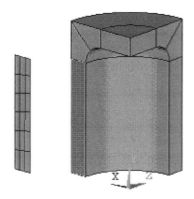

Figure 4-24 Quad mesh on lower edge.

7. Utility Menu > Plot > Volumes (To show the solid again.)

8. Main Menu > Preprocessor > Meshing > Mesh > Volume Sweep > Sweep Opts
Select the options shown below. **> OK.**

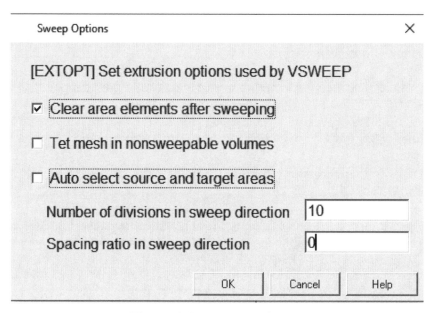

Figure 4-25 Sweep options.

9. Main Menu > Preprocessor > Meshing > Mesh > Volume Sweep > Sweep

Pick volume to be swept (Select lower portion of vessel) **> OK**

Pick source area (Select the area we just meshed with quads) **> OK** (Watch the message window lower left)

Pick target area (Select the area on the right side at 90 deg rotation) **> OK**

This process can be repeated to develop the remainder of the pressure vessel mesh. The sweep can also be performed along a line, and we can use triangular 2-D elements if a wedge type of solid element is desired.

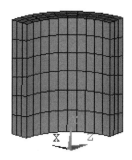

Figure 4-26 Swept mesh.

4-12 SUMMARY

This lesson concentrated on problem formulation, mesh development, solution, and results evaluation of three-dimensional problems. Free-tetrahedron, mapped-brick, and swept-brick or wedge meshes were developed and employed in the solution of the pressure vessel problem introduced in Lesson 3. Solution and post processing these kinds of problems was also illustrated.

To speed up the solution of large problems, you may want to select a working directory on your local workstation to avoid network data transfers during solution. Also investigate alternate solution methods. **Solution > Analysis Type > Sol'n Controls > Sol'n Options**.

Many 3-D problems do not have planes of symmetry and must be considered in their entirety. An example of such a part is shown below.

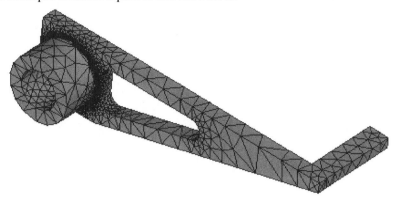

Figure 4-27 Lever model.

The part in the preceding figure is also an example where one might want to use a combined mesh of brick and tetrahedron elements. It might be desirable to map mesh the cylindrical component with bricks and the remaining portion with tets and wedges.

4-13 PROBLEMS

Use 3-D modeling methods to solve the following problems from Lesson 2:

4-1 Problem 2-1

4-2 Problem 2-2

4-3 Problem 2-3

4-4 Problem 2-4

4-5 Problem 2-5

4-6 Problem 2-6

4-7 Problem 2-7

4-8 The lever arm of Figure 4-27 is restrained in all DOF on the interior surface of the cylindrical bore and has a vertical load on the short segment of the 'L' shaped surface. Apply the total load as a series of equal concentrated forces on the nodes of the short segment. The overall length is 5 inches and the bore ID is 0.75 inches. Otherwise, select your own dimensions, materials and loads. Determine the values and location of the maximum deflection and the maximum stresses. Is failure likely to occur if the material is a ductile steel? A brittle cast iron?

4-9 Use tetrahedron solid modeling methods to solve the stress concentration problem of Lesson 2, Tutorial 2A. The model can be brought in as an IGES file, or you can use ANSYS geometry building commands to extrude an area to form a solid. Compare the results with those from Tutorial 2A and comment upon the relative ease and effectiveness of each approach.

4-10 Same as 4-9 but use 8-node bricks, then 20-node bricks.

Lesson 5

Beams

5-1 OVERVIEW

Lesson 5 treats problems in structural analysis where the most suitable modeling choice is a **beam** element. Like the truss and the plate, the beam is a **structural element**, and like the truss element it is a **line element**. Its formulation is based upon the engineering study of the behavior of beams as practical structural members. The lesson discusses

♦ Creating text file FEM beam models.

♦ Modeling 2-D and 3-D problems and evaluating the results.

5-2 INTRODUCTION

When long, slender bars are subjected to planar moments they bend in such a way that maximum compression is produced at one extreme surface and maximum tension at the other. Somewhere in between the strain is zero, and the strains and stresses vary linearly from this **neutral axis**. Such behavior is an example of pure bending.

Figure 5-1 Beam bending by moments.

The situation may not be much different when **transverse forces** cause bending, and we can determine the state of deformation by keeping track of the **displacement** of the neutral axis and its **slope**. Thus in finite element modeling of beams, we need to utilize **slope variables** in addition to the displacement variables used in previous lessons.

Figure 5-2 shows a simply supported structure composed of three beam elements carrying a point force, a concentrated moment and a distributed load. The deflected shape, displacement variables, and slope variables are also shown.

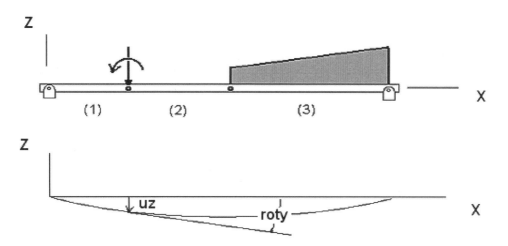

Figure 5-2 Three-element model of beam.

ANSYS provides the engineer with elastic beam elements for **two- and three-dimensional modeling**. The flexural inertias of the beam cross section provide stiffness to resist bending loadings; its cross sectional area gives it stiffness to resist axial, torsional and shearing loads.

5-3 TUTORIAL 5A – CANTILEVER BEAM

Objective: Determine the **end deflection** and **root bending stress** of the **steel** cantilever beam with **50 lbf** load shown.

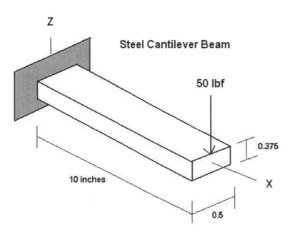

Of the several beam elements available, the ANSYS beam element **beam188** is used for modeling in this example. The ten-inch long beam is represented by **ten elements** connecting **11 nodes** along the global X-axis. Cantilever boundary conditions at the left end prevent all translational and rotational motions.

Figure 5-3 Cantilever beam.

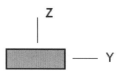

Figure 5-4 Cross Section.

The problem is formulated using a text file incorporating a **consistent set of units**, pounds force and inches in this case. The computed results will have deflections in inches, slopes in radians, shear forces in pounds, moments in inch-pounds, and stresses in psi.

```
/FILNAM,Tutorial5A
/title, 10 Element, Cantilever Beam
/prep7

!List of Nodes
n,  1,  0.0,  0.0, 0.0      ! Node 1 is located at (0.0, 0.0, 0.0) inches
n,  2,  1.0,  0.0, 0.0
n,  3,  2.0,  0.0, 0.0
n,  4,  3.0,  0.0, 0.0
n,  5,  4.0,  0.0, 0.0
n,  6,  5.0,  0.0, 0.0
n,  7,  6.0,  0.0, 0.0
n,  8,  7.0,  0.0, 0.0
n,  9,  8.0,  0.0, 0.0
n, 10,  9.0,  0.0, 0.0
n, 11, 10.0,  0.0, 0.0

n, 12, 0.0, 0.0, 5.0        !Node 12 locates the element y-z plane
                           !See figure in text.
et, 1, beam188            !Element type; number 1 is beam188

keyopt, 1, 3, 3           !Keyopt(3) = 3 for element type 1
!Key option number 3 gives a cubic bending shape variation along length

keyopt, 1, 4, 2           !Keyopt(4) = 2 for element type 1
!Key option number 4 outputs the direct shear stress results

sectype,1, beam, rect    !Cross section number 1 is rectangular
secdata, 0.5, 0.375      !Cross section base = 0.5, height = 0.375

!Material Properties
mp, ex, 1, 3.e7          ! Elastic modulus for material number 1 in psi
mp, prxy, 1, 0.3        ! Poisson's ratio

!List of elements and nodes they connect
en,  1, 1,  2, 12        ! Element Number 1 connects nodes 1 & 2
en,  2, 2,  3, 12        ! Dummy node Node 12 orients the cross section
en,  3, 3,  4, 12
en,  4, 4,  5, 12
en,  5, 5,  6, 12
en,  6, 6,  7, 12
en,  7, 7,  8, 12
en,  8, 8,  9, 12
en,  9, 9, 10, 12
en, 10, 10, 11, 12

!Displacement Boundary Conditions
d, 1, all, 0.0          ! All DOF at node 1 are zero
d, 12, all, 0.0         ! All DOF at dummy node 12 are zero

!Applied Force
f, 11, fz, -50.         ! Force at node 11 in negative z-direction is 50 lbf.
```

```
/pnum, elem, 1        ! Plot element numbers
eplot                 ! Plot the elements
finish
```

The ANSYS 3D beam element **BEAM188** is used in modeling this problem. A typical element located in a **global coordinate system XYZ** is shown below. It connects two nodes **I** and **J,** and has its **local** or **element x-axis** defined by these two nodes. The **local y** and **z** axes are aligned with the principal cross section flexural inertia planes.

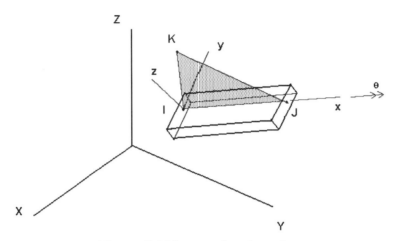

Figure 5-5 Three node orientation.

A third node, **K,** is used to define the **local x-z plane** which is one of the principal planes of bending of the beam. Node K can be another node in the model or a dummy node (with all DOF set to zero) that is used just for orientation purposes. The beam cross sectional properties are defined with respect to the local axis system. The values for **Izz** and **Iyy** correspond to the orientation specified by the three nodes. Check your work carefully. It's easy to get Izz and Iyy reversed in planning your analysis.

1. Start ANSYS, etc., read the input file using **File > Read Input from . . .**

PlotCtrls > Symbols > All Applied BCs > OK and **SOLVE**

Plot the deformed shape and examine the **computed deflections** and **slopes**. To get X-Z plane select ⌷ Bottom View

Figure 5-6 Ten-element model.

List the beam downward **displacements**.

2. Main Menu > General Postproc > Read Results > First Set.

Main Menu > General Postproc > List Results > Nodal Solution > DOF Solution > Displacement vector sum > OK

```
PRINT U   NODAL SOLUTION PER NODE

  ***** POST1 NODAL DEGREE OF FREEDOM LISTING *****

  LOAD STEP=    1  SUBSTEP=     1
   TIME=   1.0000     LOAD CASE=   0

  THE FOLLOWING DEGREE OF FREEDOM RESULTS ARE IN THE GLOBAL COORDINATE
SYSTEM

     NODE      UX          UY          UZ           USUM
        1   0.0000      0.0000       0.0000       0.0000
        2   0.0000      0.0000      -0.36936E-02 0.36936E-02
        3   0.0000      0.0000      -0.14214E-01 0.14214E-01
        4   0.0000      0.0000      -0.30802E-01 0.30802E-01
        5   0.0000      0.0000      -0.52700E-01 0.52700E-01
        6   0.0000      0.0000      -0.79150E-01 0.79150E-01
        7   0.0000      0.0000      -0.10939      0.10939
        8   0.0000      0.0000      -0.14267      0.14267
        9   0.0000      0.0000      -0.17822      0.17822
       10   0.0000      0.0000      -0.21529      0.21529
       11   0.0000      0.0000      -0.25311      0.25311

MAXIMUM ABSOLUTE VALUES
NODE         0           0           11           11
VALUE    0.0000      0.0000      -0.25311      0.25311
```

List the beam **slopes**.

3. Main Menu > General Postproc > List Results > Nodal Solution > DOF Solution > Rotation vector sum > OK

```
PRINT ROT  NODAL SOLUTION PER NODE

  ***** POST1 NODAL DEGREE OF FREEDOM LISTING *****

  LOAD STEP=    1  SUBSTEP=     1
   TIME=   1.0000     LOAD CASE=   0

  THE FOLLOWING DEGREE OF FREEDOM RESULTS ARE IN THE GLOBAL COORDINATE
SYSTEM

     NODE     ROTX         ROTY        ROTZ         RSUM
        1   0.0000      0.0000       0.0000       0.0000
        2   0.0000      0.72059E-02  0.0000       0.72059E-02
        3   0.0000      0.13653E-01  0.0000       0.13653E-01
```

```
    4    0.0000     0.19342E-01    0.0000     0.19342E-01
    5    0.0000     0.24273E-01    0.0000     0.24273E-01
    6    0.0000     0.28444E-01    0.0000     0.28444E-01
    7    0.0000     0.31858E-01    0.0000     0.31858E-01
    8    0.0000     0.34513E-01    0.0000     0.34513E-01
    9    0.0000     0.36409E-01    0.0000     0.36409E-01
   10    0.0000     0.37547E-01    0.0000     0.37547E-01
   11    0.0000     0.37926E-01    0.0000     0.37926E-01

MAXIMUM ABSOLUTE VALUES
NODE         0          11            0           11
VALUE    0.0000     0.37926E-01    0.0000     0.37926E-01
```

The **maximum deflection UZ** and **maximum slope ROTY** occur at node 11 at the free end as expected.

The maximum values -0.2531 inch and -0.03792 radians agree with results you can calculate from solid mechanics beam theory. You can confirm these results with a simple hand calculation. Now examine the **computed bending stress**:

4. Main Menu > General Postproc > List Results > Element Solution > Stress > X-Component of stress > OK

```
PRINT S    ELEMENT SOLUTION PER ELEMENT

  STRESSES AT BEAM SECTION NODAL POINTS

ELEMENT =      1  SECTION ID =        1

ELEMENT NODE = 1

     SEC NODE        SXX            SXZ            SXY
        1         -42667.        -66.667      -0.49978E-12
        3         -42667.        -66.667      -0.24961E-12
       13         0.62016E-12    -466.67      -0.28051E-12
       11         0.26466E-12    -466.67       0.33216E-12
        5         -42667.        -66.667      -0.16652E-11
       15         0.97566E-12    -466.67      -0.29033E-12
       23          42667.        -66.667      -0.67484E-12
       21          42667.        -66.667      -0.18065E-11
       25          42667.        -66.667       0.24457E-11

     Max=          42667.       0.24457E-11   -66.667

     Min=         -42667.      -0.18065E-11   -466.67

ELEMENT NODE = 2

     SEC NODE        SXX            SXZ            SXY
        1         -38400.        -66.667      -0.49978E-12
        3         -38400.        -66.667      -0.24961E-12
       13        -0.84753E-12    -466.67      -0.28051E-12
       11        -0.20836E-11    -466.67       0.33216E-12
        5         -38400.        -66.667      -0.16652E-11
       15         0.38858E-12    -466.67      -0.29033E-12
```

```
   23          38400.        -66.667       -0.67484E-12
   21          38400.        -66.667       -0.18065E-11
   25          38400.        -66.667        0.24457E-11

Max=          38400.         0.24457E-11   -66.667

Min=         -38400.        -0.18065E-11   -466.67
```
. . . .

The stresses in each beam element are given in the form shown. The data above is for the first element connecting nodes 1 and 2, the one at the cantilever support. **SXX** is the stress in the X-Direction, perpendicular to the cross section. The values are given at **cross section** nodal points 1, 3, 13, . . . , 25. These points are nodes created by the SECTYPE and SECDATA commands when APDL computes the geometric properties of the cross section.

5. Main Menu > Preprocessor > Sections > Plot Section > Show section mesh? > Node numbers > OK

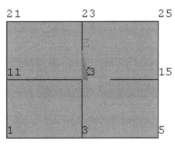

Figure 5-7 Beam cross section.

Eight-node elements are used to describe the 2D geometry of the cross section and to compute the properties of interest in the analysis of beams, A, Iyy, Izz, etc. Only corner nodes are shown in the plot. Nodes 1, 3, and 5 are on the **lower side of the beam** and are in **compression**. Nodes 21, 23, and 25 are on **the top side and are in tension**.

The **maximum bending stress** of **42,667 psi** occurs in **beam element 1** at the support, and the value agrees with what you would compute using elementary beam theory, Mc/I without considering any stress concentration effects.

Stresses SXZ and SXY are the shear stresses in the beam and are also shown. In a cantilever beam with a rectangular cross section, a small shear stress, SXZ, is produced by the transverse load at the end; you can compute it using solid mechanics theory also to check the results above. It should be zero at top and bottom surfaces but the mesh on the cross section is not dense enough to approach that result. The cross section mesh density can be increased for greater accuracy.

The displacement and stress solution for this problem is solved just as accurately with **one 10-inch long element** connecting nodes at each end of the beam. The deformed shape plot, however, would show a straight line connecting the two nodes because the plotting software does not use computed slope information. If examining the deformed shape is important or if the spatial mass distribution needs to be accurately represented for vibration problems, use several nodes along the length as described above, otherwise two nodes will do the job, but the plotted results may look funny.

Axial stiffness is included in the beam188 element formulation, so we could add an axial force to the above problem and compute the axial stress and deformation as well. Note, however, that in the **linear approach** discussed here, the **axial and bending deformations are uncoupled**. That is, the presence of an axial stress does not influence the bending stiffness. If the bending deformation is small, this approach is usually completely satisfactory.

Figure 5-8 Transverse and axial loads.

To illustrate this, add an axial force of 1000 lbf to the end of the beam and solve the problem again.

Add this line to the text file   **f, 11, fx, 1000**

6. Utility Menu > File > Clear & Start new > OK > Yes

7. Utility Menu > File > Read Input from . . . (Read in the file with the axial load added.) List the displacements.

8. Main Menu > General Postproc > Read Results > First Set.
Main Menu > General Postproc > List Results > Nodal Solution > DOF Solution > Displacement vector sum > OK

```
   THE FOLLOWING DEGREE OF FREEDOM RESULTS ARE IN THE GLOBAL COORDINATE
SYSTEM

      NODE      UX            UY            UZ            USUM
         1   0.0000        0.0000        0.0000        0.0000
         2   0.17778E-03   0.0000       -0.36936E-02   0.36979E-02
         3   0.35556E-03   0.0000       -0.14214E-01   0.14218E-01
         4   0.53333E-03   0.0000       -0.30802E-01   0.30807E-01
         5   0.71111E-03   0.0000       -0.52700E-01   0.52705E-01
         6   0.88889E-03   0.0000       -0.79150E-01   0.79155E-01
         7   0.10667E-02   0.0000       -0.10939       0.10940
         8   0.12444E-02   0.0000       -0.14267       0.14267
         9   0.14222E-02   0.0000       -0.17822       0.17822
        10   0.16000E-02   0.0000       -0.21529       0.21529
        11   0.17778E-02   0.0000       -0.25311       0.25312

MAXIMUM ABSOLUTE VALUES
NODE         11            0             11            11
VALUE     0.17778E-02   0.0000        -0.25311       0.25312
```

The deflections show **axial deformations** now, but the bending displacements and slopes are exactly the same as before indicating **no interaction** between the axial and bending loads.

List the stresses. (Results for the first element are shown.)

9. Main Menu > General Postproc > List Results > Element Solution > Stress > X-Component of stress > OK

```
PRINT S    ELEMENT SOLUTION PER ELEMENT

 STRESSES AT BEAM SECTION NODAL POINTS

 ELEMENT =        1  SECTION ID =        1

 ELEMENT NODE = 1

     SEC NODE         SXX             SXZ             SXY
         1         -37333.         -66.667      -0.49978E-12
         3         -37333.         -66.667      -0.24961E-12
        13          5333.3         -466.67      -0.28051E-12
        11          5333.3         -466.67       0.33216E-12
         5         -37333.         -66.667      -0.16652E-11
        15          5333.3         -466.67      -0.29033E-12
        23         48000.          -66.667      -0.67484E-12
        21         48000.          -66.667      -0.18065E-11
        25         48000.          -66.667       0.24457E-11

    Max=          48000.        0.24457E-11    -66.667

    Min=         -37333.       -0.18065E-11    -466.67
```

The added axial load produces a direct **tensile stress** (P/A) of 5333 psi that is combined with the bending stresses to give maximum and minimum values at the cantilever support of 48,000 psi on the top of the beam and -37,333 psi on the bottom.

For problems with **large deformations**, the nonlinear coupling of axial stress and bending stiffness must be considered and ANSYS **nonlinear solution** options employed. What is large and what is small for a given problem? Some previous experience and/or numerical experimentation can help answer that question.

5-4 2-D FRAME

As another example of beam element modeling with ANSYS, we find the stress and deflection distribution in the two-dimensional frame of Figure 5-6. This model is contrived to include most of the possible situations one might encounter, including a downward force (**7000 lbf**), a concentrated moment (**1000 in-lbf**) and a wind load acting on the left side. The wind loading is developed from the load per area acting on the side of the structure that is converted to a line loading in load per unit length. It varies from **10 lbf/in** at the top to **zero lbf/in** at the bottom.

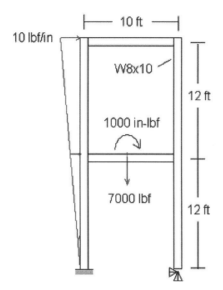

Figure 5-9 Frame.

The frame is constructed of **W8x10** standard shape beams. The area of the cross section is **2.96 sq. in**. and the flexural inertia is **30.8 in⁴**. The total section depth is **7.89 in**. The frame is rigidly attached at its left support but allowed to pivot at its right support. If the I beams are made of A36 steel (Sy = 36,000 psi), will any yielding occur?

5-5 TUTORIAL 5B – 2-D FRAME

We use a text file to define the model, and perform the solution and postprocessing interactively. ANSYS element **beam3**, not supported any longer in interactive mode, is at this writing functional when referenced in a text file. Note how the node dimensions are **converted** from **feet** to **inches** in the text file.

1. Start ANSYS and enter the following data using **Read Input From ...** Then **Plot > Elements**, etc.

```
/FILNAM,Tutorial5B
/title, Tutorial 5B – 2-D Frame

/prep7

et, 1, beam3     ! Element type; no.1 is beam3

!Material Properties
mp, ex, 1, 3.e7          ! Elastic modulus
mp, prxy, 1, 0.3         ! Poisson's ratio
mp, dens, 1, 0.283/386.  ! Mass density


!List of Nodes
n, 1, 0.0,     0.0       ! Node 1 is located at (0.0, 0.0)inches
```

```
! Convert from feet to inches for consistent units.
n,   2, 10.0*12,   0.0
n,   3,  0.0,      12.0*12
n,   4,  5.0*12,   12.0*12
n,   5, 10.0*12,   12.0*12
n,   6,  0.0,      24.0*12
n,   7, 10.0*12,   24.0*12
```

! **'Real'** constant set 1 for W8x10 I beam. **Out dated format but still supported**
! Area, Izz (flexural Inertia), height 'h' as in sigma = Mc/I, c = h/2
! A = 2.96 sq.in., Izz = 30.8 in^4, h = 7.89 inch

```
r, 1, 2.96, 30.8, 7.89
```

```
!List of elements and nodes they connect
en,  1,  1,  3         ! Element Number 1 connects nodes 1 & 3
en,  2,  2,  5
en,  3,  3,  4
en,  4,  4,  5
en,  5,  3,  6
en,  6,  5,  7
en,  7,  6,  7
```

```
!Displacement Boundary Conditions
d, 1, ux, 0.0          ! Displacement at node 1 in
x-dir is zero
d, 1, uy, 0.0          ! Displacement at node 1 in
y-dir is zero
d, 1, rotz, 0.0        ! Rotation about z axis at
node 1 is zero

d, 2, ux, 0.0
d, 2, uy, 0.0
```

```
!Applied Loadings
f, 4, fy, -7000. ! Force at node 4 in negative y-
direction is 7000 lbf.
f, 4, mz, -1000  ! Moment about Z axis is -1000
in-lbf

sfbeam, 1, 1, pres, 0, 5   ! Surface Force on
beam 1 varies from 0 to 5
sfbeam, 5, 1, pres, 5, 10

acel, 0, 386., 0      ! acceleration of gravity

finish
```

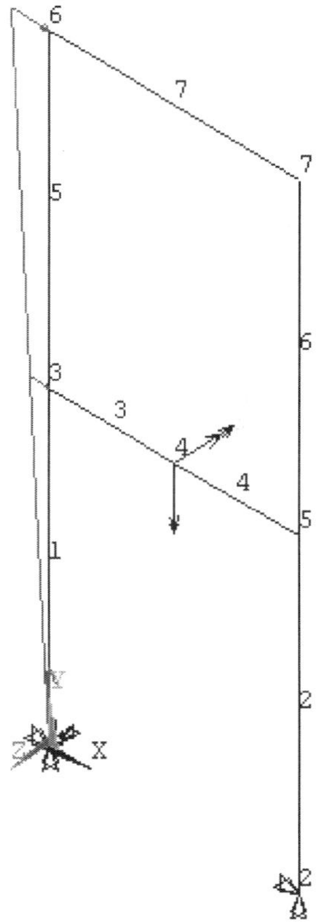

Figure 5-10 Frame model.

Note that boundary conditions must be supplied for the **slope variables** and that here **line loads** are applied to the beam elements using the **sfbeam** command. Each beam element has a local coordinate axis with local x directed from **node I** (first named node) to **node J** (second named node) and with y and z axes according to the right hand rule with z

directed toward the viewer. The direction of the surface loading requires some pre-analysis planning and/or interactive experimentation with the model to get things right. To view the loads and element coordinate systems.

2. Utility Menu > PlotCtrls > Symbols > [/PSF] Surface Load Symbols (set to **Pressures**) and **Show pres and convect as** (set to **Arrows**).

Utility Menu > PlotCtrls > Symbols > ESYS Element Coordinate sys (set to **ON**).

To view the concentrated moment and the slope boundary condition, select **ISO** in the **Pan, Zoom, Rotate** options.

3. Main Menu > Solution > Solve > Current LS > OK

When plotting the deformed shape, notice that even though the beam element deformed shape has been correctly computed, as a plotting convenience the nodes are connected by straight lines. If you want to visualize the beam element shape, include a few extra nodes and elements between the points of the beam connections.

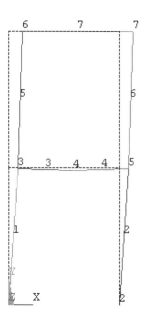

Figure 5-11 Deformed frame.

4. Main Menu > General Postproc > Read Results > First Set.
Main Menu > General Postprocess > Plot Results > Deformed Shape > Def + undeformed. Now list the nodal deflection values.

5. Main Menu > General Postproc > List Results > Nodal Solution > DOF Solution > Displacement vector sum > OK

```
PRINT DOF  NODAL SOLUTION PER NODE

 ***** POST1 NODAL DEGREE OF FREEDOM LISTING *****

LOAD STEP=     1  SUBSTEP=     1
  TIME=    1.0000      LOAD CASE=   0

THE FOLLOWING DEGREE OF FREEDOM RESULTS ARE IN GLOBAL COORDINATES

    NODE     UX           UY          UZ         USUM
       1  0.0000       0.0000       0.0000      0.0000
       2  0.0000       0.0000       0.0000      0.0000
       3  0.49528     -0.35297E-02  0.0000      0.49529
       4  0.49537     -0.14501      0.0000      0.51616
       5  0.49547     -0.87344E-02  0.0000      0.49555
       6  0.73410     -0.31099E-02  0.0000      0.73411
       7  0.73316     -0.95128E-02  0.0000      0.73322
```

```
MAXIMUM ABSOLUTE VALUES
NODE           6              4              0              6
VALUE     0.73410       -0.14501       0.0000        0.73411
```

As an alternative to the method used in the Tutorial 5A to examine stress results, here we will create an Element Table. Because the direct and bending stresses combine to produce different results on the top and bottom of the beam and also produce different results at each end of the beam, we will create four element table entries using interactive commands.

6. Main Menu > General Postprocess > Element Table > Define Table > Add

Enter a label you choose such as **SmaxI** and scroll down to find **By sequence num**.

Select **NMISC** and enter **1** (to the right of NMISC) > **Apply**.

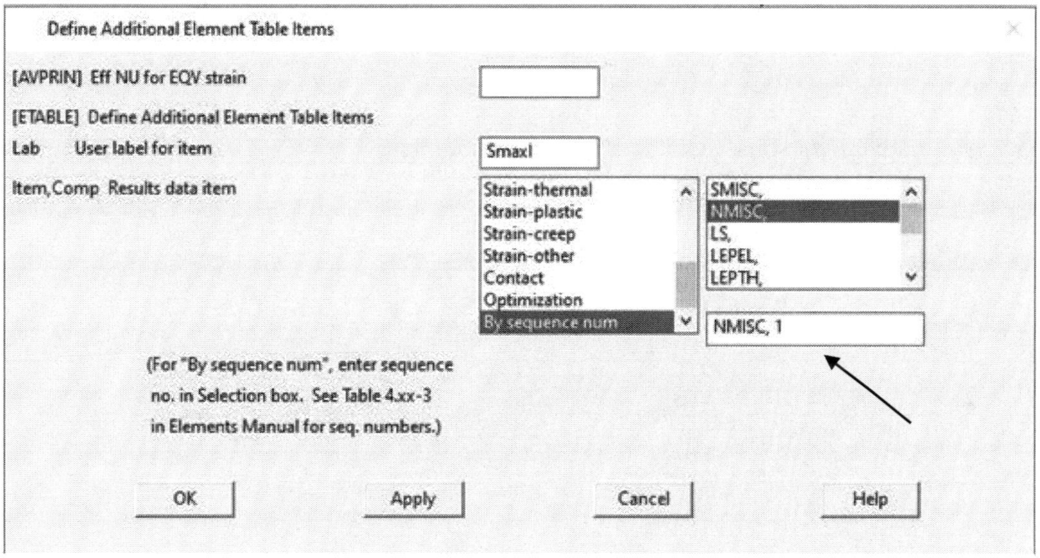

Figure 5-12 Define element table quantities.

Repeat for

Label **SminI, By sequence num, NMISC, 2 > Apply**

Label **SmaxJ, By sequence num, NMISC, 3 > Apply**

Label **SminJ, By sequence num, NMISC, 4 > OK > Close Element Table Data**

This sets up tabular data for Smax and Smin at nodes I and J for each element.

7. Main Menu > General Postprocess > Element Table > List Elem Table (Click on the first four items SMAXI, SMINI, etc.) > **OK**

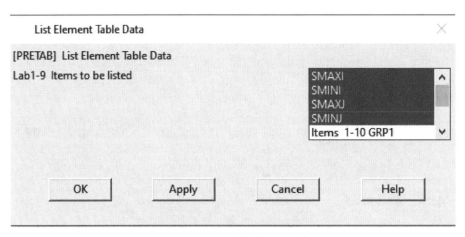

Figure 5-13 Table data to list.

The resulting element stresses are shown in the table below.

```
PRINT ELEMENT TABLE ITEMS PER ELEMENT

  ***** POST1 ELEMENT TABLE LISTING *****

    STAT     CURRENT      CURRENT      CURRENT      CURRENT
    ELEM      SMAXI        SMINI        SMAXJ        SMINJ
       1     10119.      -11631.       2525.3      -3955.3
       2    -1840.0      -1840.0       8431.5     -12030.
       3      515.64     -419.78      18113.      -18017.
       4     18241.      -18145.      18439.      -18343.
       5      2839.7      -2705.5      1099.5       -883.83
       6      7978.0      -8343.0      4545.1      -4828.7
       7       756.35     -1227.0      4451.6      -4922.2

MINIMUM VALUES
ELEM           2            4            5            4
VALUE     -1840.0      -18145.       1099.5      -18343.

MAXIMUM VALUES
ELEM           4            3            4            5
VALUE     18241.       -419.78      18439.       -883.83
```

A maximum stress of 18,439 psi occurs in element 4, but there will be no yielding in the frame for the material yield stress is 36 kpsi.

You can also plot each of these table data items for a visual indication of the stress distribution in the frame. See below.

Figure 5-14 Plot of SMAXI.

8. Main Menu > General Postprocess > Element Table > Plot Elem Table (Select the one to plot.) **> OK**

5-6 BEAM MODELS IN 3-D

When a beam element is incorporated in a 3-dimensional model, the full 3-D flexibility of the beam must be considered. It can have **axial deformation, torsional deformation**, and **bending deformations** in **two principal bending planes.** Each beam of the model must be positioned in space to reflect the proper orientation of the element cross section.

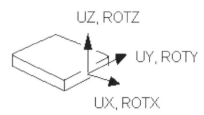

Figure 5-15 Beam DOF.

The correct orientation can be defined by specifying the angular rotation of the element about its longitudinal axis or by employing three node points to define a principal plane of bending for the element. The next tutorial uses the approach with 3 nodes.

5-7 TUTORIAL 5C – 'L' BEAM

Objective: Find the stresses and deflections of a simple 'L'-shaped aluminum beam with one end cantilevered and a point load at the other end.

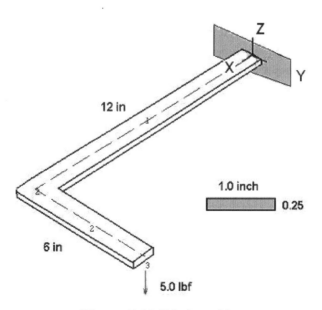

Figure 5-16 'L'-shaped beam.

The ANSYS 3D beam element **BEAM4** is used in modeling this problem. A typical element located in a **global coordinate system XYZ** is shown below. It connects two nodes **I** and **J**, and has its **local** or **element x-axis** defined by these two nodes. The **local y** and **z** axes are aligned with the principal cross section flexural inertia planes.

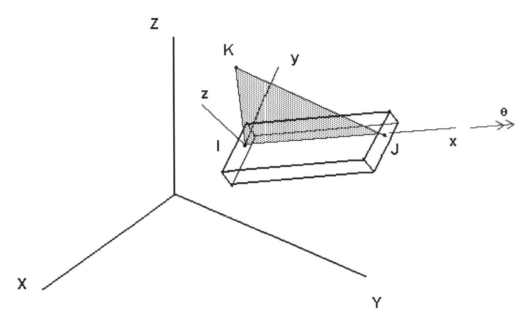

Figure 5-17 Global and local axes.

A third node, **K**, is used to define the **local x-z plane** which is one of the principal planes of bending of the beam. Node K can be another node in the model or a dummy node (with all DOF set to zero) that is used just for orientation purposes. The beam cross sectional properties are defined using an ANSYS 'Real Constants' set. The values for **Izz** and **Iyy** entered into the program **must correspond** to the orientation specified by the three nodes. Check your work carefully. It's easy to get Izz and Iyy reversed.

The Global axis (X,Y,Z) and local axis (x,y,z) definitions for the two elements of the simple problem of this tutorial are illustrated below

Important: The geometric property constant **Ixx** is the **torsional stiffness constant** for the cross section. For circular sections Ixx is equal to the polar moment of inertia (Iyy + Izz). For non-circular sections the torsional stiffness constant is **not equal** to the polar moment of inertia (see torsion of non-circular section in a solid mechanics reference). If no value is entered for Ixx, however, ANSYS will compute the torsional stiffness constant as Iyy + Izz which is correct for circular sections but not correct for non-circular sections.

The text file below defines the simple 'L' – Beam ANSYS model.

```
/FILNAM,Tutorial5C
/title, 3D Beam Sample Problem - An 'L' - Beam
/prep7

n, 1,     0.0,  0.0,  0.0    ! Node 1 is located at (X=0.0, Y=0.0, Z=0.0)
n, 2,  12.0,  0.0, 0.0
n, 3,  12.0,  6.0,  0.0
n, 4,  12.0,  0.0,  4.0


et, 1, beam188                    ! Element type; no.1 is 3D beam188

!Material Properties
mp, ex, 1, 1.e7                ! Elastic modulus, psi
mp, prxy, 1, 0.3              ! Poisson's ratio

keyopt, 1, 3, 3          !Keyopt(3) = 3 for element type 1
!   Key option number 3 gives proper bending moment variation along
length

keyopt, 1, 4, 2          !Keyopt(4) = 2 for element type 1
!   Key option number 4 =2 turns on transverse + torsion shear output

sectype,1, beam, rect    !Cross section number 1 is rectangular
secdata, 1.0, 0.25          !Cross section base = 1.0, height = 0.25

! element connection and orientation
en, 1, 1, 2, 4      ! Element #1 connects nodes 1 & 2, and
                                ! uses node 4 to define the element local x-z
plane.

en, 2, 2, 3, 4      ! Element #2 connects nodes 2 & 3, and
                                ! uses node 1 to define the element local x-z
plane.

! Displacement boundary conditions
d, 1, ux, 0.                    ! Displacement at node 1 in x-dir is zero
d, 1, uy, 0.
d, 1, uz, 0.
d, 1, rotx, 0.                  ! Rotation at node 1 about x-axis is zero
d, 1, roty, 0.
d, 1, rotz, 0.

! (We could have used d, 1, all, 0.0 to define all root restraints.)

! Applied force
f, 3, fz, -5.0                  ! Force at node 3 in the negative y-direction
finish

/solu                          ! Select static load solution
antype, static
solve
save
finish

/post1
```

1. Read the problem definition file into ANSYS, and once it's solved, plot the deformed shape.

The deformed plus undeformed beam is shown in the figure below. Nodes were only used at the ends of the 'L' segments so the deformed shape shows a straight-line connection between nodes. The calculated deformations are correct, however. You can check the results by hand using the superposition or energy methods.

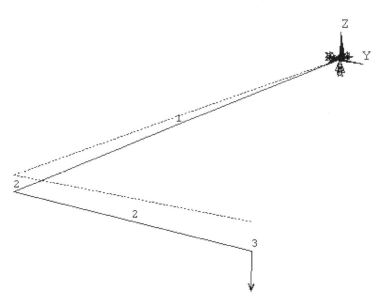

Figure 5-18 Deformed two-element beam.

The rotation boundary conditions **ROTX, ROTY, ROTZ** are indicated by double-headed arrows, positive by the right-hand rule. They overlay the displacement conditions **UX, UY, UZ** at the cantilever support; on screen the two conditions are shown in different colors. Now list the deflection, slope and stress results.

2. Main Menu > General Postproc > Read Results > First Set.
Main Menu > General Postproc > List Results > Nodal Solution > DOF Solution > Displacement vector sum > OK

```
PRINT U    NODAL SOLUTION PER NODE

 ***** POST1 NODAL DEGREE OF FREEDOM LISTING *****

 LOAD STEP=     1  SUBSTEP=     1
   TIME=    1.0000     LOAD CASE=   0

 THE FOLLOWING DEGREE OF FREEDOM RESULTS ARE IN THE GLOBAL COORDINATE
SYSTEM

   NODE    UX          UY          UZ          USUM
      1  0.0000      0.0000      0.0000      0.0000
```

```
      2 -0.22272E-16 0.83206E-14-0.22126      0.22126
      3 -0.94058E-14 0.83206E-14-0.37273      0.37273

MAXIMUM ABSOLUTE VALUES
NODE            3            3            3            3
VALUE  -0.94058E-14 0.83206E-14-0.37273      0.37273
```

3. Main Menu > General Postproc > List Results > Nodal Solution > DOF Solution > Rotation vector sum > OK

```
PRINT ROT   NODAL SOLUTION PER NODE

  ***** POST1 NODAL DEGREE OF FREEDOM LISTING *****

 LOAD STEP=      1  SUBSTEP=       1
  TIME=    1.0000      LOAD CASE=    0

 THE FOLLOWING DEGREE OF FREEDOM RESULTS ARE IN THE GLOBAL COORDINATE
SYSTEM

    NODE      ROTX         ROTY         ROTZ         RSUM
      1   0.0000       0.0000       0.0000       0.0000
      2 -0.20631E-01 0.27648E-01 0.13678E-14 0.34497E-01
      3 -0.27543E-01 0.27648E-01 0.16266E-14 0.39026E-01

MAXIMUM ABSOLUTE VALUES
NODE            3            2            3            3
VALUE  -0.27543E-01 0.27648E-01 0.16266E-14 0.39026E-01
```

4. Main Menu > General Postproc > List Results > Element Solution > Stress > X-Component of stress > OK

```
PRINT S    ELEMENT SOLUTION PER ELEMENT

  STRESSES AT BEAM SECTION NODAL POINTS

ELEMENT =         1  SECTION ID =          1

ELEMENT NODE = 1

    SEC NODE        SXX           SXZ           SXY
        1        -5760.0        126.16        -1045.7
        3        -5760.0        -5.0000       -1855.9
       13     -0.16226E-009    -35.000    -0.24857E-010
       11      0.34426E-008    796.06      0.94111E-011
        5        -5760.0       -136.16       -1045.7
       15     -0.37671E-008    -866.06    -0.81855E-011
       23         5760.0        -5.0000        1855.9
       21         5760.0        126.16         1045.7
       25         5760.0       -136.16         1045.7

    Max=         5760.0        796.06         1855.9

    Min=        -5760.0       -866.06        -1855.9

ELEMENT NODE = 2
```

```
        SEC NODE        SXX             SXZ             SXY
           1        -0.95960E-004      126.16          -1045.7
           3        -0.95964E-004     -5.0000          -1855.9
          13        -0.16204E-009     -35.000           0.75460E-011
          11         0.39999E-008      796.06           0.14115E-010
           5        -0.95968E-004     -136.16          -1045.7
          15        -0.43240E-008     -866.06          -0.36380E-011
          23         0.95964E-004     -5.0000           1855.9
          21         0.95968E-004      126.16           1045.7
          25         0.95959E-004     -136.16           1045.7

        Max=         0.95968E-004      796.06           1855.9

        Min=        -0.95968E-004     -866.06          -1855.9

    STRESSES AT BEAM SECTION NODAL POINTS

 ELEMENT =          2   SECTION ID =            1

 ELEMENT NODE = 2

        SEC NODE        SXX             SXZ             SXY
           1        -2880.0           -5.0000           0.30591E-008
           3        -2880.0           -5.0000           0.56813E-008
          13        -0.13126E-012     -35.000           0.33763E-009
          11         0.39374E-008     -35.000           0.48267E-010
           5        -2880.0           -5.0000           0.30589E-008
          15        -0.39377E-008     -35.000           0.48211E-010
          23         2880.0           -5.0000          -0.50061E-008
          21         2880.0           -5.0000          -0.29626E-008
          25         2880.0           -5.0000          -0.29625E-008

        Max=         2880.0           -5.0000           0.56813E-008

        Min=        -2880.0           -35.000          -0.50061E-008

 ELEMENT NODE = 3

        SEC NODE        SXX             SXZ             SXY
           1        -0.47950E-004     -5.0000          -0.45620E-007
           3        -0.47949E-004     -5.0000          -0.81000E-007
          13         0.24680E-013     -35.000          -0.41042E-010
          11        -0.35167E-009     -35.000          -0.58294E-011
           5        -0.47949E-004     -5.0000          -0.45620E-007
          15         0.35167E-009     -35.000          -0.58848E-011
          23         0.47949E-004     -5.0000           0.80918E-007
          21         0.47949E-004     -5.0000           0.45608E-007
          25         0.47950E-004     -5.0000           0.45608E-007

        Max=         0.47950E-004     -5.0000           0.80918E-007

        Min=        -0.47950E-004     -35.000          -0.81000E-007
```

The computed deflection and stress values can be checked easily by hand since the simple 'L'-beam is statically determinate. The deformations are in the global axis system and are caused by bending of element 2; by bending and twisting of element 1.

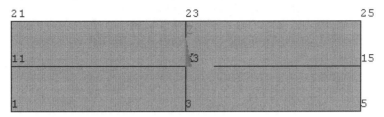

Figure 5-19 Element cross section mesh.

In the local coordinate system for each element, SXX is the bending stress in each, positive on the top side and negative on the bottom; SXZ is the shear due to the transverse shear **plus or minus** the torsional shear; SXY is the torsional shear on top and bottom nodes. Note for element 1 the difference between SXZ for cross section nodes 11 and 15. Also note the equal and opposite values for SXY at nodes 3 and 23. Beam element 2 is in bending and shear only.

An element table can also be used to facilitate evaluation of computed results. Search on beam188 in the online help and find the output table definitions for stresses, etc.

5-8 SUMMARY

ANSYS modeling of two and three-dimensional beam bending problems has been presented in this lesson and tutorials given to provide hands-on experience with this kind of analysis. The engineering theory of beam bending is the basis for the elements utilized here, and angular degrees of freedom (ROTX, etc.) are introduced for the first time since these variables are required for the calculation of the beam element neutral-axis slope.

Transverse loads that cause bending also cause shearing deformations that may become important for short beams or beams that have high bending stress allowables but are relatively weak in shear such as beams made of wood. The real property set for beams allows specification of the effective shear deformation area so that this behavior can be modeled when necessary.

Because they connect two node points, beam and truss elements are called 'line' elements in ANSYS. The Shell (plate) elements of the next lesson are the two-dimensional or surface equivalents of beams, connecting three or four nodes in a plane, they are loaded by forces transverse to their surface and experience bending in ways similar to beams.

5-9 PROBLEMS

Use 2-D beam models to find the maximum deflection (mm or inches) and bending stress (N/m^2 or psi) for the rectangular cross section single span beams of Problems 5-1 through 5-5 below. Use metric system units, **M,** or British system units, **B,** for problem formulation. Compare your computed results with those that you calculate from elementary beam theory.

Length	$L = 3$ m	9.8 ft.
Elastic modulus	$E = 2.\text{E}11$ N/m^2	2.9E7 psi
Cross section Base	$b = 50$ mm	1.97 in
Cross section Height	$h = 150$ mm	5.9 in.
Distributed load	$w = 30$ kN/m	2055 lbf/ft
Point load	$P = 25$ kN	5620 lbf
Point moment	$M = 12$ kN-m	8852 ft-lbf

5-1 Simply supported beam (a) point load P at center, (b) uniformly distributed load w.

5-2 Fixed (cantilevered) at one end, free at the other, (a) point load P at free end, (b) uniformly distributed load w, (c) point moment M at free end, (d) linearly distributed load varying from w at the fixed end to zero at the free end.

5-3 Fixed (cantilevered) at both ends, (a) point load P at center, (b) uniformly distributed load w.

5-4 Fixed (cantilevered) at one end, simply supported at the other, (a) point load P at center, (b) uniformly distributed load w.

5-5 Find the maximum deflection and stress in the stepped steel cantilevered beam below with a point load on the free end. The segments are of equal length and the larger inertia is twice the smaller. Select your own material, geometric, and load values. Check the maximum bending stress with a simple hand calculation. The exact end deflection should be somewhere between that for a single beam of inertia I and that for one with an inertia $2I$.

Figure P5-5

5-6 A stepped steel shaft with equal length segments of 16 inches, and diameter cross sections of 1.5 inch, 1 inch is loaded with a 1000 lbf force in the middle and considered restrained from deflection and slope by the bearings at either end. (a) Compute the mid-span deflection and moment. (b) Compute the magnitude and location of the maximum deflection and moment.

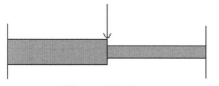

Figure P5-6

5-7 Find the maximum stress and deflection of the 2-D frame shown if it has both columns fixed at ground level. The geometric, material, and distributed load parameters are the same as for the beam of Problem 5-1. There is a side load of *w* and also a uniformly distributed downward load *w* (not shown) on each horizontal beam.

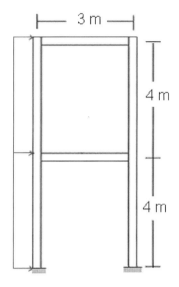

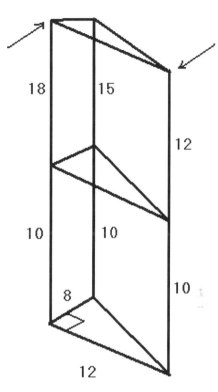

Figure P5-7   **Figure** P5-10

5-8 Determine the maximum stress in the 'L'-beam of Figure 5-11 if a 10 lbf force is applied at the free end in the global Z direction in addition to the 5.0 lbf Y-direction force shown. Confirm the ANSYS result with a hand calculation.

5-9 Create a 3-dimensional model of a building frame using two frames like that of Problem 5-7, the second displaced 3 m in the Z-direction from the first. The second frame is loaded like the first and connected to it by with beams at the 4 and 8 m levels. The connecting beams (same properties as the others) have distributed loads 2*w* N/m. Find magnitude and location of the maximum stress and the maximum displacement.

5-10 The tower in the figure has a right triangle base and is loaded by two 5000 lbf forces parallel to the direction of the 8 ft side of the triangle. Find the maximum displacement and stress if the tower is made from steel pipe with a 4.5 inch OD and a wall thickness of 0.337 inches. The three legs are firmly anchored at ground level, so the three base triangle lines do not need to be elements of the model. Use BEAM4 or PIPE16 (needs only two nodes to define) elements.

NOTES:

Lesson 6

Shells

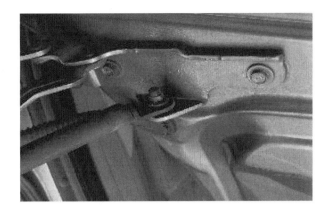

6-1 OVERVIEW

Shell (or plate) elements can resist loads in all directions and become extremely versatile modeling tools for thin planar and curved surfaces. The exercises below demonstrate

♦ Modeling flat plates with ANSYS shell elements.

♦ Evaluating shell element solution accuracy.

♦ Solving problems involving thin walled structures.

6-2 INTRODUCTION

Shell elements behave as the two-dimensional equivalents of beams. They connect three or four node points in a plane and when loaded by forces transverse to their surfaces, experience bending about the X-axis and about the Y-axis (Fig 6-1). These deformations cause bending stresses that vary linearly through the shell thickness. Transverse loads can also cause twisting as well as bending. The **middle surface** of the plate is a plane of zero bending strain as is the neutral axis plane of a beam.

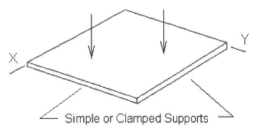

Figure 6-1 Shell

We will use the terms plate and shell interchangeably in this discussion. As with beams, if the plate thickness is a tenth or less than the length dimension, pure bending dominates, but for greater thickness to length ratios, through-the-thickness shearing deformation begins to become important as well as bending.

Four and eight-node quad shell elements are provided in the ANSYS element library. They support bending, shearing and in-plane loadings. Like beams, the in-plane or membrane loadings are not coupled with the bending loadings for linear analysis.

Plate element theory does not naturally supply stiffness to resist a torque applied normal to the shell surface, but this deficiency is overcome by introducing a physically reasonable stiffness parameter in the mathematical formulation in order to provide elements that resist loading in all six degrees of freedom at each node.

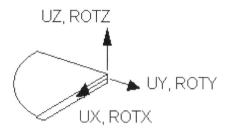

Figure 6-2 Shell degrees of freedom

6-3 SQUARE PLATE

As an example of the use of ANSYS shell elements, consider a **10 inch** by **10 inch 304 steel plate** that **is 0.1 inch thick** and has a **1.0 lbf force** applied at its center. The plate is **simply supported** on all sides.

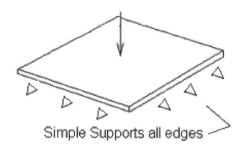

Figure 6-3 Simply supported, point-loaded plate.

6-4 TUTORIAL 6A - SQUARE PLATE

Objective. Determine the deflection at the load point as a function of the number of elements used per side to model the plate and find the von Mises stress distribution.

The plate has two planes of symmetry, so we model just one quadrant and apply appropriate boundary conditions. For starters, use the ANSYS simple **four-node** element, **SHELL181** for the model.

Start ANSYS and select the working directory for this problem then select the element.

1. Main Menu > Preprocessor > Element Type > Add/Edit/Delete > Add > Structural Shell > 3D   4 Node 181 > Close

2. Material > Props > Material Models. Enter material properties for 304 steel, **E = 2.8E7, prxy = 0.29 > Close**

Specify the **plate Thickness.** The **Sections > Shell > Lay-Up** sequence can be used to build models of a **composite shell.** Here we use it for a **single layer shell.**

3. Main Menu > Preprocessor > Sections > Shell > Lay-Up > Add/Edit > Thickness
Enter **0.1 > OK.** Take the default **Material ID = 1.**

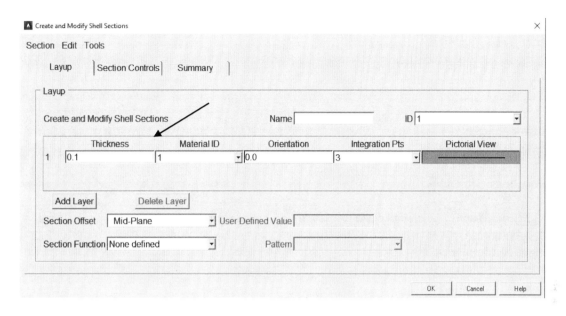

Figure 6-4 Enter plate thickness.

Create a 10 inch x 10 inch area in the X-Y plane.

**4. Main Menu > Preprocessor > Modeling > Create > Areas > Rectangle > By 2
Corners >** Enter **WP X = -5.0, WP Y = -5.0, Width = 10.0, Height = 10.0 > OK**

For meshing first set the size controls so that 4 divisions per line are created. Then mesh
the area to create 4 elements per edge.

**5. Main Menu > Preprocessor > Meshing > Size Controls > Manual Size > Lines >
All Lines > NDIV Number of element divisions > Enter 4 > OK**

6. Main Menu > Preprocessor > Meshing > Mesh > Areas > Mapped > 3 or 4 Sided
(Select the 10 x 10 square)

Now apply **boundary conditions**
and **loads**. The global axis system is
located at the center of the plate; the
edges are **simply supported**
meaning they cannot move in the Z-
direction. The simply supported edge
also means no rotation in a direction
perpendicular to the edge.

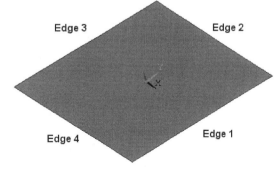

Figure 6-5 Simply supported plate.

Apply the load to the **Node** at the origin of coordinates.

7. Main Menu > Preprocessor > Loads > Define Loads > Apply > Structural > Force/Moment > On Node > Select the Node at the center of the plate. For FZ enter –1.0 > OK

Boundary Conditions on Edges 1 and 3

8. Main Menu > Preprocessor > Loads > Define Loads > Apply > Structural > Displacement > On Lines > Select Edge 1 and Edge 3 > Set UZ = 0 > OK Then repeat and **set ROTX = 0.**

Now constrain edges 2 and 4

9. Main Menu > Preprocessor > Loads > Define Loads > Apply > Structural > Displacement > On Lines > Select Edge 2 and Edge 4 > Set UZ = 0 > OK Then repeat and **set ROTY = 0.**

Solve the equations

10. Main Menu > Solution > Solve > Current LS > OK

A solution error occurs because all **rigid body degrees of freedom** have not been restrained.

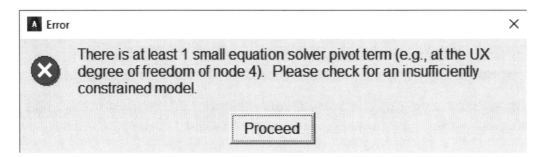

Figure 6-6 Insufficiently constrained model.

To restrain rigid body motion, we will set edge displacements in X and Y to zero.

Edges 1, 2, 3 and 4

11. Main Menu > Preprocessor > Loads > Define Loads > Apply > Structural > Displacement > On Lines > Select Edges 1, 2, 3 and 4 > Set UX = 0 > OK Then repeat and **set UY = 0.**

Orange double headed arrows indicate rotation constraints.

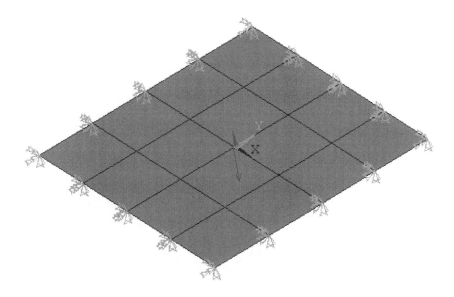

Figure 6-7 Model showing mesh, load and boundary conditions.

Solve the equations.

12. Main Menu > Solution > Solve > Current LS > OK

Post processing

13. Main Menu > General Postproc > Read Results > First Set

Main Menu > General Postproc > Plot Results > Deformed Shape > Def shape only.
(Control right mouse button to rotate view.)

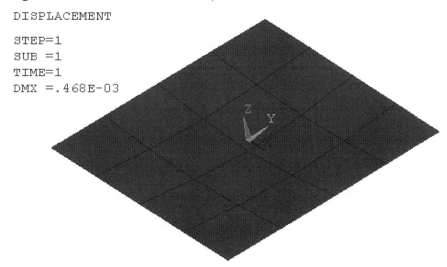

Figure 6-8 Deformed shape.

Plot the **contours of vertical displacement** over the plate.

14. Main Menu > General Postproc > PLOT RESULTS > Contour Plot > Nodal Solu > DOF Solution > Z-Component of displacement > OK

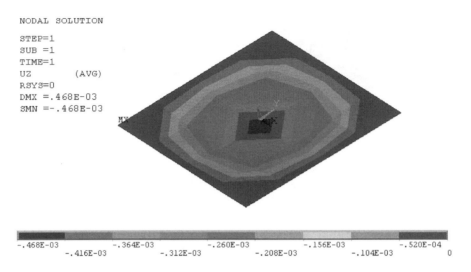

Figure 6-9 UZ contours.

List values or use the results viewer to determine the deflection at the **center** of the plate.

15. Main Menu > General Postproc > LIST RESULTS > Nodal Solu > DOF Solution > Z-Component of displacement > OK

The maximum deflection at the center of the plate is **-0.46797E-03 inches.** The **theoretical solution** to this problem gives a deflection value of **-0.452E-3 inches**. Thus, our **mesh is not fine enough** using this element to give an accurate deflection solution. As verification, plot the calculated **von Mises Stress.** SEQV is the combination of the **SX** and **SY bending stresses** at the **upper** or the **lower surface** of the plate.

16. Main Menu > General Postproc > Plot Results > Contour Plot > Element Solu > Stress > von Mises stress > OK   (Repeat but select the **Nodal Solution.**)

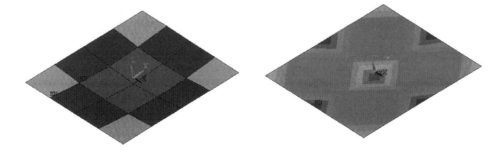

Figure 6-10 Element Solution / Nodal Solution for von Mises stress contours.

The discontinuities in stress contours also indicate an inaccurate stress solution for this mesh and element. To improve the accuracy, remove the applied load, clear the mesh (**Preprocessor > Meshing > Clear > Areas**), remesh with 8 elements per edge, and reapply the load. Then repeat with 16 elements per edge, etc. The deflection results are shown below.

Elements per edge	Deflection
4	0.468E-3
8	0.458E-3
16	0.456E-3
Theory	0.452E-3

We observe that the 16-elements-per-edge mesh produces deflection with an **error of less than 1 percent** when compared to thin plate theory. Also notice that the maximum displacement is **converging from above**, meaning that the shell elements are **softer** than the theoretical solution we are using for comparison.

Contours of vertical deflection UZ and von Mises stress for the 16 by 16 mesh are shown below. The stress contours are now much smoother in comparison with the first mesh. However, the element solution stress plots show discontinuities near the point-load singularity at the center of the plate (as we would expect) and elsewhere. A still finer mesh is needed for more accurate stress results.

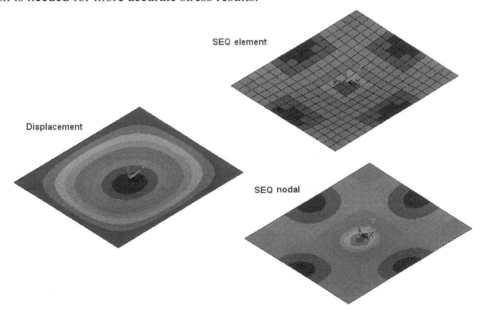

Figure 6-11 UZ and von Mises stress contours for 16 x 16 mesh.

Since the stresses are proportional to the strains (the spatial derivatives with respect to *x* and *y* of the displacements), the **stresses** for a given mesh generally have **greater error than the displacements**. In other words, the displacements are more accurate.

For large problems it may be important to make use of symmetry and employ only a quadrant of the plate. The displacement normal to a line of symmetry must be zero and the slope of the plate along the line of symmetry must be zero.

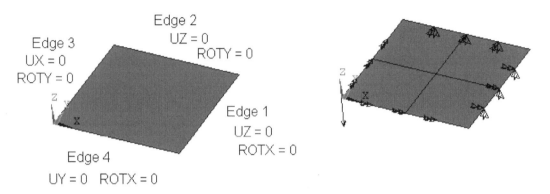

Figure 6-12 Quadrant of simply supported plate.

To obtain yet greater accuracy for this problem, increase the mesh density or switch to the **more accurate eight-node SHELL281** element, or do **both**. Because the point load produces a singular point in the stress distribution, we cannot calculate an accurate stress value at the load application point no matter the mesh density. Stresses elsewhere are accurate, however. The calculated deflections are also accurate.

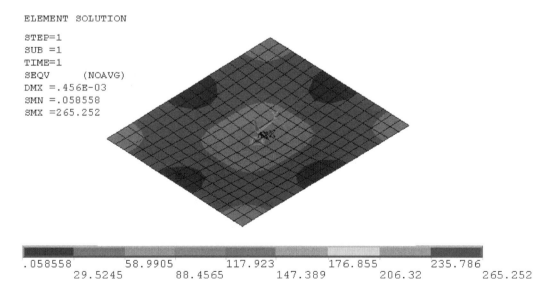

Figure 6-13 NOAVG Von Mises stress using 8 node shell281 elements.

The ANSYS shell elements are capable of modeling a wide range of practical problems as shown in the next example.

6-5 TUTORIAL 6B - CHANNEL BEAM

Objective: Find the deformations and stresses in a **cantilevered thin walled channel** subject to an **end load**. The 6 x 2 x 0.2 inch thick steel channel shown below is 30 inches long and has a total load of 1.0 lbf equally divided among the four corner points on the right end as shown.

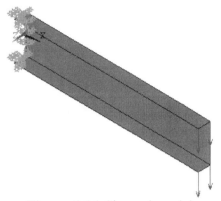

Figure 6-14 Channel model

The following text file is used to create the geometry.

```
/FILNAM,Tutorial6B
/title, Channel geometry

/prep7
! Create geometry
k, 1,  0.,  -3., 0.        ! Keypoint 1 is at -3.0, 0.0
k, 2,  0.,   3., 0.
k, 3,  0.,  -3., 2.
k, 4,  0.,   3., 2.
k, 5, 30.,  -3., 0.
k, 6, 30.,   3., 0.
k, 7, 30.,  -3., 2.
k, 8, 30.,   3., 2.

L, 1, 2                    ! Line connecting keypoints 1 & 2
L, 2, 4
L, 3, 1
L, 4, 8
L, 2, 6
L, 6, 8
L, 6, 5
L, 7, 5
L, 1, 5
L, 3, 7

AL, 1, 5, 7, 9            ! Area defined by lines 1, 5, 7, 9
```

```
AL, 2, 5, 6, 4
AL, 3, 9, 8, 10
```

Start ANSYS, select the working directory for this problem and select a job name.

1. Read the text file to create the geometry **File > Read Input from …**

We will use the eight-node **SHELL281** element for this example.

2. Main Menu > Preprocessor > Element Type > Add/Edit/Delete > Add > Structural Shell > Elastic 8 Node 281 > Close

3. Enter **E = 3.E7** and **prxy = 0.3** material properties for steel. **Close.**

Specify the plate thickness.

4. Main Menu > Preprocessor > Sections > Shell > Lay-Up > Thickness Enter **0.2 > OK.** Take the default **Material ID = 1.**

Use size controls and mapped meshing to create **2** elements along each flange, **4** elements through the web, and **10** elements along the length.

For the flanges

5. Main Menu > Preprocessor > Meshing > Size Controls > Manual Size > Lines > Picked Lines > (Select the short edges of upper and lower flanges) **NDIV Number of element divisions > Enter 2 > OK**

Repeat for the web height and length edges then mesh all three areas.

6. Main Menu > Preprocessor > Meshing > Mesh > Areas > Mapped > 3 or 4 sided > Pick All

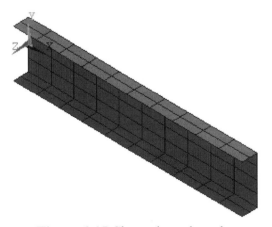

Figure 6-15 Channel quad mesh.

7. Restrain all DOF along the lines on the left end and apply **Fy = -0.25 lbf** to the four keypoints on the right end.

8. Solve and open the **Postprocessor** to plot the deformed shape and the von Mises stresses as shown below. The maximum downward deflection is about 0.2E-3 inch.

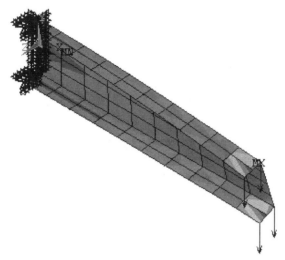

Figure 6-16 Displaced shape and von Mises stress contours.

The observed twisting occurs because the resultant end load is not applied through the **shear center** of the thin-walled cross section. The point loads cause the high stresses on the right end, and if you zoom in you see that the mesh may need refinement at points away from the ends to obtain better stress accuracy.

6-6 ORTHOTROPIC PLATES

Consider a 20 x 20 x 0.1 inch plate made of an **orthotropic material** with properties

$$Ex = 21.6e6 \text{ psi}$$
$$Ey = 1.54e6 \text{ psi}$$
$$Ez = 1.54e6 \text{ psi}$$
$$PRXY = 0.253$$
$$PRYZ = 0.421$$
$$PRXZ = 0.253$$
$$Gxy = 6E5 \text{ psi}$$
$$Gyz = 4.66E5 \text{ psi}$$
$$Gxz = 6E5 \text{ psi}$$

Use the modeling tools to create a 20 x 20 inch area.

Figure 6-17 Plate area 20 x 20.

Select 8 node Solid183 Element, with options Plane Stress with thickness.

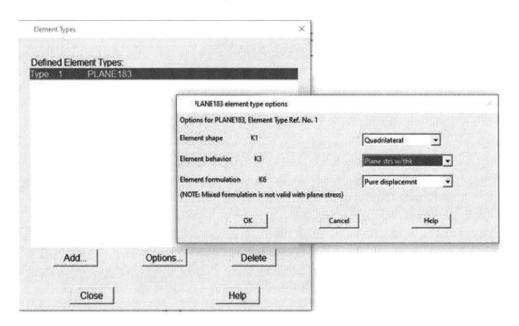

Figure 6-18 Solid 183, plane stress with thickness.

Real constants > Add

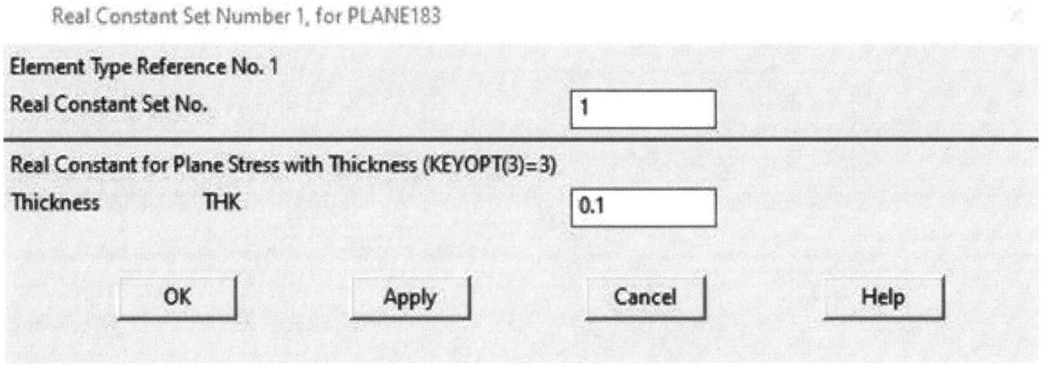

Figure 6-19 Enter real constant thickness.

Select **Material Props > Material Models > Linear Orthotropic Material**. Properties for the z direction are unknown but assumed to be the same as those for the y-direction.

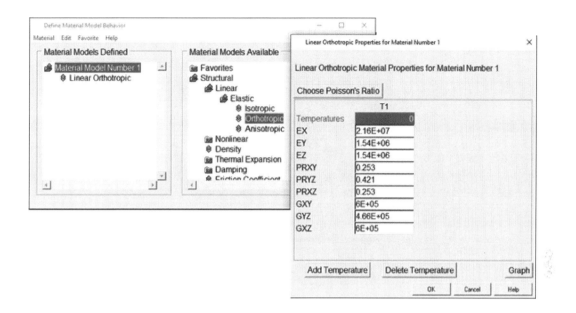

Figure 6-20 Enter Orthotropic material properties.

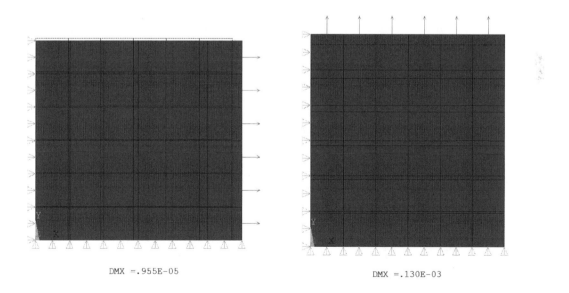

DMX =.955E-05   DMX =.130E-03

Figure 6-21 Displacement response to a -10 psi pressure, first in X dir, then in Y dir.

DMX is the maximum deformation in the figures above, different in each direction for the same load value due to the orthotropic material behavior.

6-7 COMPOSITE PLATES

Composite parts are created by bonding together thin orthotropic layers as shown in the figure below. The number of layers and directions of principal stiffness can be selected to produce desired specific deflection and stress behavior.

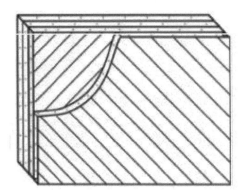

Figure 6-22 Composite Plate (Kokcharov - Wikipedia).

It is not unusual for many layers to be used to produce a desired structural response. We use a two layer composite example here to illustrate the finite element approach to this type of problem.

Consider a **5 x 12 inch** laminated plate made of the **same orthotropic material** as the plate of the previous example. The plate has **two 0.005 inch thick layers (plies)**. One ply is oriented at **0 degrees** with respect to the longitudinal axis; the other is orientated at **45 degrees**.

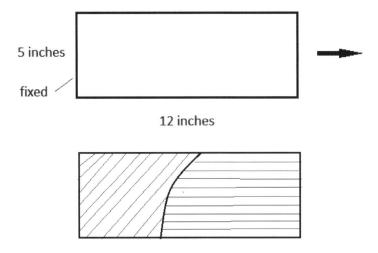

Figure 6-23 Two ply plate, cut away.

The plate is completely restrained at one end and loaded at the other with a 5 psi tensile load (apply a negative pressure -5psi to create this load). Use the SHELL281 eight node shell (plate) element. > **Options** (Note Bottom 1st top last)

SHELL281 element type options   ×

Options for SHELL281, Element Type Ref. No. 1

Element stiffness	K1	Bending and membrane ▼
Curved shell formulation	K5	Advanced ▼
Storage of layer data	K8	Bottom 1st top last ▼
User Thickness option	K9	No UTHICK routine ▼
Normal stress (Sz) output	K10	Not modified ▼
Default element x axis	K11	First parametric ▼

OK   Cancel   Help

Figure 6-24 Select Shell281 element.

To model the composite nature of the problem use the shell281 layup option.

Preprocessor > Sections > Shell > Lay-up > Add/Edit (Enter layer 1 then Add Layer)

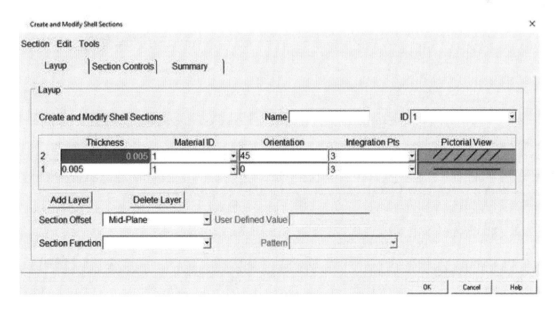

Figure 6-25 Shell281 lay-up specification.

Mesh and apply boundary conditions, left edge completely fixed, right edge **tensile load** of 5 psi. and **Solve** the current load step LS.

Post processor > **Read** the Results > **Plot** the deformed shape and undeformed shape.

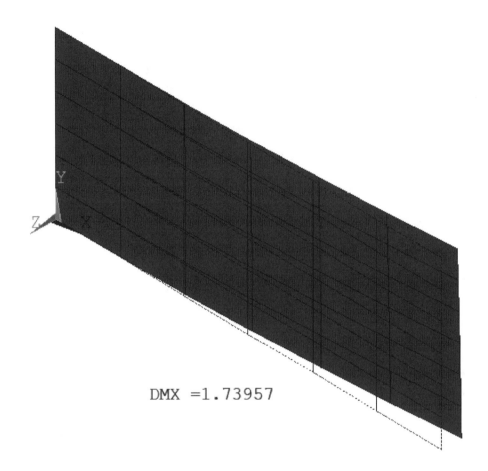

Figure 6-26 Deformed and undeformed image of composite plate.

It is interesting to note that because of the presence of the 45 degree ply, a **X-direction load** of 5psi in the x-y plane of the plate produces a large out-of-plane deformation and a twisting in addition to the small extension in the direction of the pressure.

6-8 SUMMARY

Thin plate or shell elements find applications in many practical situations as illustrated by the tutorials in this lesson. They are particularly useful for the analysis of parts such as structural angle brackets, automotive hoods or trunk lid attachments and similar thin components with fairly complex shape. In these cases shells may offer a modeling and analysis advantage over comparable models that use solid elements. Examples of orthotropic and composite materials used in plate applications are also presented to illustrate the practical application of such important materials in engineering practice.

6-9 PROBLEMS

6-1 Duplicate the analysis of Tutorial 6A but change the boundary conditions to **fixed** on **all edges** and the loading to a **uniform pressure** of 1.0 psi. Compare your results to results calculated from published solutions for thin plate deformation and stress.

6-2 Repeat Tutorial 6B but use a T, H, L, or I thin-walled section. Load symmetric sections off center so as to cause twisting.

6-3 Use a solid modeler to develop the geometry of an angle bracket or hood hinge attachment. Import the middle surface of the solid model into ANSYS, then formulate suitable loading and boundary conditions for analysis with ANSYS shell elements.

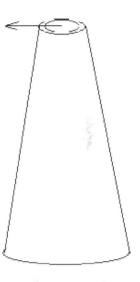

6-4 Solve any of the tutorials or problems from this lesson with solid tetrahedron or brick elements and compare the results of the two kinds of modeling in terms of modeling ease, computational effort and interpretation of results.

6-5 A vertical thin-walled conical tube is completely fixed at its base and is loaded by a uniformly distributed horizontal force at its top. Use SHELL281 elements for modeling and determine the transverse deflection at the top and the maximum von Mises stress at mid height if it is made of steel. Select your own dimensions, material properties, and loads.

Figure P6-5

6-6 Repeat Problem 6-5 but use the orthotropic material from example 6-5.

6-7 Repeat Problem 6-5 but use a composite material together with a four ply 0, 90, 0,90 lay-up. Change the cone into a cylinder.

6-8 Repeat Problem 6-7 but use a composite material together with a four ply 45, -45, 45, -45 lay-up.

NOTES:

Lesson 7

Heat Transfer & Thermal Stress

7-1 OVERVIEW

Heretofore we have considered only problems in structural and continuum mechanics. ANSYS capabilities also include modeling of problems involving the behavior of thermal, electric and magnetic systems. Lesson 7 demonstrates:

♦ Determining temperature distributions in conduction/convection problems.

♦ Using temperature distributions to find thermal stresses.

7-2 INTRODUCTON

Linear thermal conduction/convection problems are formulated using the finite element approach with temperature as the single degree-of-freedom variable at each node in the mesh and with the material conduction properties used to form the thermal 'stiffness matrix' to be solved.

The ANSYS library includes one-, two-, and three-dimensional thermal elements for modeling problems of interest to practicing engineers. The temperature distributions found at each thermal analysis node can be stored for use with the equivalent structural model for use as nodal 'inputs' for finding stresses caused by temperature changes.

7-3 HEAT TRANSFER

Triangular and quadrilateral thermal elements are provided to perform thermal analysis for planar and axisymmetric models, while tetrahedron and brick shaped elements are available for three-dimensional problems. A correspondence in the number of nodes and their locations between thermal and structural models allows switching between the two types of analysis without creating a new mesh.

We start by solving a conduction/convection problem defined in two-dimensions. The theoretical solution for this problem is available in the literature for comparison.

7-4 TUTORIAL 7A – TEMPERATURE DISTRIBUTION IN A CYLINDER

Objective: We wish to compute the temperature distribution in the **cross section** of a **long** steel cylinder with inner radius **5 inches** and outer radius **10 inches**. The interior of the cylinder is kept at **75 deg F**, and heat is lost on the exterior by convection to a fluid whose temperature is **40 deg F**. The **convection coefficient is 2.E-4 BTU/sec-sq.in-F** and the **thermal conductivity** for steel is taken to be **8.09 E-4 BTU/sec-in-F**.

Figure 7-1 Cylinder with internal temperature and exterior convection.

1. Start ANSYS and assign a job name to the project. **Run Interactive > set working directory** and **jobname.**

2. Main Menu > Preferences > Select **Thermal > OK** (This is a thermal problem.)

Select the eight-node thermal quad element. (Scroll down the list of elements to find thermal.)

3. Main Menu > Preprocessor > Element Type > Add/Edit/Delete > Add > Thermal Solid > 8 node 77 > OK > Close

Set the material properties.

4. Main Menu > Preprocessor > Material Props > Material Models > Thermal > Conductivity > Isotropic > Enter **KXX = 8.09 E-4** (BTU/sec-in-F) **> OK > Close window.**

Create the geometry.

5. Recognizing the symmetry of the problem, a quadrant of a section through the cylinder is created using ANSYS area creation tools. **Main Menu > Preprocessor > Modeling > Create > Areas > Circle > Partial Annulus**

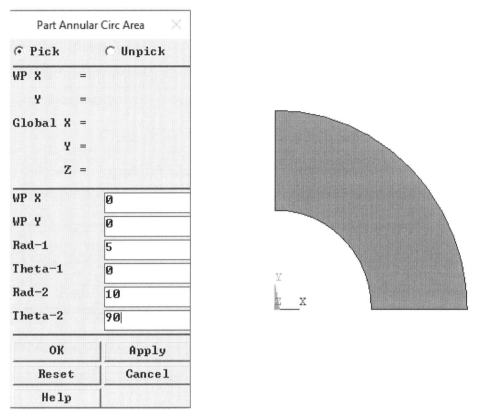

Figure 7-2 Quadrant of cylinder.

6. Set the number of line divisions to 5 radially and 8 circumferentially and use mapped meshing, 3 or 4 sided areas, to create the 5 by 8 mesh shown.

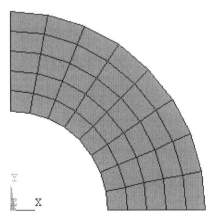

Figure 7-3 Quad mesh.

7. Main Menu > Preprocessor > Loads > Define Loads > Apply > Thermal > Temperatures > On Lines Select the interior boundary and set the temperature to **75**.

8. Main Menu > Preprocessor > Loads > Define Loads > Apply > Thermal > Convection > On Lines

Select the line defining the outer surface; set the convection **film coefficient to 2e-4** and the **bulk temp to 40**. (Leave VALJ and VAL2J fields blank.) **OK**

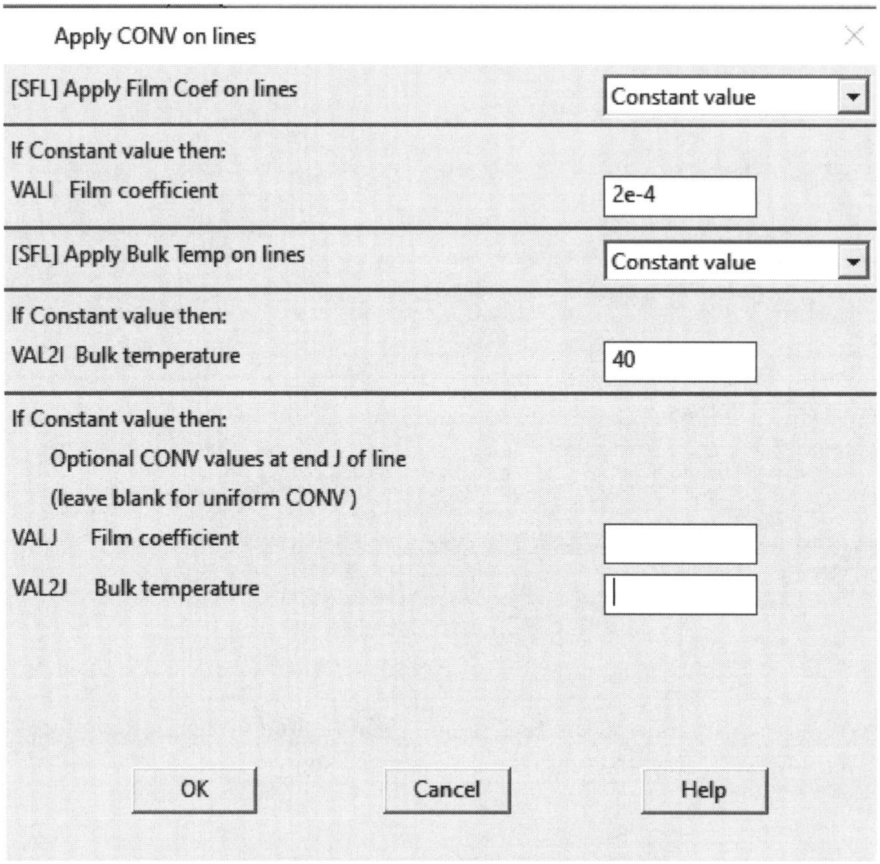

Apply CONV on lines   ✕

[SFL] Apply Film Coef on lines | Constant value ▼ |

If Constant value then:
VALI   Film coefficient | 2e-4 |

[SFL] Apply Bulk Temp on lines | Constant value ▼ |

If Constant value then:
VAL2I   Bulk temperature | 40 |

If Constant value then:
    Optional CONV values at end J of line
    (leave blank for uniform CONV)
VALJ   Film coefficient | |
VAL2J   Bulk temperature | |

| OK | | Cancel | | Help |

Figure 7-4 Convection boundary condition.

To account for symmetry, select the **vertical** and **horizontal lines of symmetry** and set the **heat flux** to **zero**.

9. Main Menu > Preprocessor > Loads > Define Loads > Apply > Thermal > Heat Flux > On Lines > 0 > OK

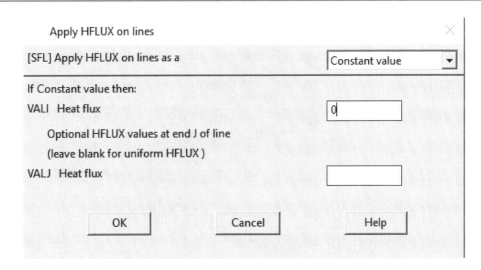

Figure 7-5 Zero flux on lines of symmetry.

Use **PlotCtrls > [/PSF] Surface Load Symbols > Convect FilmCoef > Show pres as > Arrows** to view the convection boundary conditions. Or **List > Loads > Surface > On All Lines** to view all surface boundary conditions.

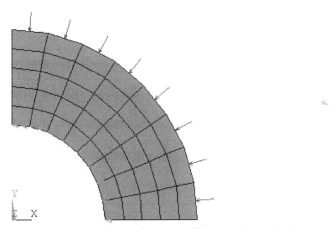

Figure 7-6 Temperature and convection boundary conditions **after solution step**.

10. Main Menu > Solution > Solve > Current LS

11. Main Menu > General Postprocessor > Plot Results > Contour Plot > Nodal Solu > DOF Solution > Nodal Temperature > OK

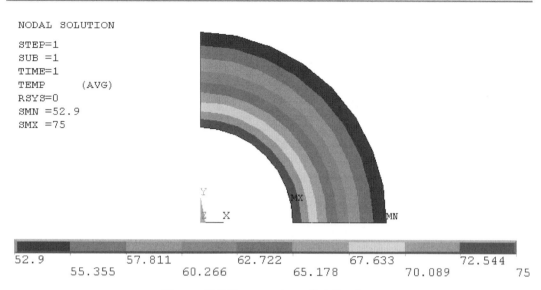

```
NODAL SOLUTION

STEP=1
SUB =1
TIME=1
TEMP      (AVG)
RSYS=0
SMN =52.9
SMX =75
```

52.9   57.811   62.722   67.633   72.544
     55.355   60.266   65.178   70.089   75

Figure 7-7 Temperature distribution.

The maximum temperature is on the interior and is the specified 75 F. On the outside wall the minimum temperature is found to be about **53 F**. These results can be verified using closed form solutions from heat transfer theory.

12. Save your work.

Because of its symmetry, this problem could have been solved with a much smaller model, say a 10 or 15 degree wedge by setting the heat flux on the cut faces to zero.

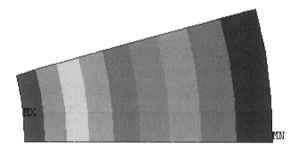

Figure 7-8 Temperature distribution for wedge model.

7-5 THERMAL STRESS

Objective: To determine the stress in an object due to a temperature change.

When a temperature change occurs in an object, its coefficient of thermal expansion causes an expansion or contraction of the object depending upon whether the temperature increases or decreases. If the object is unrestrained, there is no stress associated with this free motion. On the other hand, if the object is subject to restraint against the thermally induced movement, a stress can occur.

The finite element method provides tools for the solution of thermal stress problems. The temperature changes can be uniform throughout the body or distributed due to thermal conditions such as those in the previous tutorial.

7-6 TUTORIAL 7B – UNIFORM TEMPERATURE CHANGE

Consider a block of material (1) surrounded by a second material (2) as shown in the figure below. The composite is subjected to a temperature change so that both regions attain a uniform final temperature throughout. Find the stresses that result if the initial state is stress free. We're missing a couple of components here, but this is similar to the situation that occurs during the encapsulation of computer chips.

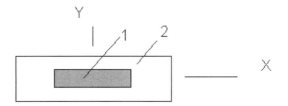

Figure 7-9 Encapsulation model.

Assume a representative geometry of about 6 x 25 mm for a plane stress model. The text file below generates the **upper right quadrant** of the geometry and includes hypothetical material properties with different elastic moduli and different CTEs, **coefficients of thermal expansion**, *alpx* for the two components.

```
/FILNAM,Tutorial7B
/title, Chip geometry

/prep7
! Create geometry
k, 1,    0,   0.        ! Keypoint 1 is at 0.0, 0.0
k, 2,   9.5,  0.
k, 3,  12.5,  0.
k, 4,    0,   1.6
k, 5,   9.5,  1.6
k, 6,  12.5,  1.6
k, 7,    0,   3.2
k, 8,   9.5,  3.2
k, 9,  12.5,  3.2
```

```
L, 1, 2                          ! Line connecting keypoints 1 & 2
L, 2, 3
L, 4, 5
L, 5, 6
L, 7, 8
L, 8, 9
L, 1, 4
L, 2, 5
L, 3, 6
L, 4, 7
L, 5, 8
L, 6, 9

AL, 1, 8, 3, 7                   ! Area defined by lines 1, 8, 3, 7
AL, 2, 9, 4, 8
AL, 3, 11, 5, 10
AL, 4, 12, 6, 11

mp, ex, 1, 3.e4                  ! Material 1 properties, including CTE
mp, prxy, 1, 0.3
mp, alpx, 1, 10.3e-6

mp, ex, 2, 2.5e3
mp, prxy, 2, 0.3
mp, alpx, 2, 100.e-6
```

1. Start ANSYS, etc. **Set preferences to Structural**. (This is a **structural problem** with a **thermal loading**.) Read in the text file defining the upper right quadrant of the geometry shown in the previous figure. Turn on **area numbering (PlotCtrls > Numbering AREA numbers > ON)** and **plot the areas**.

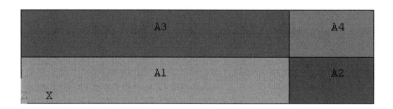

Figure 7-10 Areas.

2. Select the **8-node quad 183** structural element (plane stress, unit thickness is default).

3. Main Menu > Preprocessor > Meshing > Mesh Attributes > Picked Areas > Select area A1 > OK > Area Attributes (select area A1 to be Material 1) **> OK**

4. Repeat specifying areas **A2, A3, A4** to be Material 2.

5. Use **manual size controls** to divide the interior horizontal and vertical line segments into **12, 4, 4,** and **4 divisions**. Then **Mapped Mesh > 3 or 4 sided > Pick All**

Check to see that the material assignments are correct.

6. Utility Menu > PlotCntls > Numbering > Elem / Attribute numbering > Material Numbers > OK

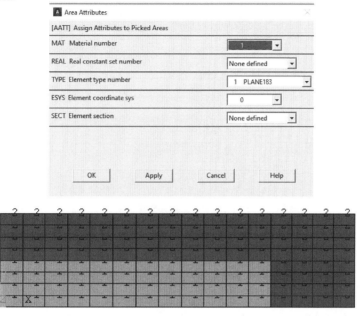

Figure 7-11 Area Attributes and Two-material mesh.

7. Apply **symmetric displacement boundary conditions** along bottom edge (UY = 0) and left edge (UX = 0).

Apply a **uniform temperature change** of **-150 deg C** representing a cool down during the encapsulation operation from the initial stress-free state.

8. Main Menu > Preprocessor > Loads > Define Loads > Apply > Structural > Temperature > On Areas > Pick All > Enter -150 > OK

9. Save your work, **Solve,** and then go to the **Postprocessor.**

10. Read Results > First Set. Plot Results The **deformed shape** (note shrinkage) and the **first principal stress, S1,** are shown in the two figures below.

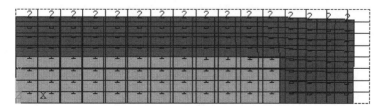

Figure 7-12 Deformed shape.

Figure 7-13 First principal stress.

The stress plot shows the **singularity** at the interior right angle of the composite object. Examine the other principal stresses. The largest principal stress in magnitude is **S3**.

7-7 TUTORIAL 7C – THERMAL STRESS

Objective: In this exercise we wish to first compute the temperature distribution in the vessel below, Figure 7-14, and then compute the thermal stresses that result after the thermal changes assuming an initial stress-free state.

A steel cylinder with inner radius **5 inches** and outer radius **10 inches** is 40 inches long and has spherical end caps. The interior of the cylinder is kept at **75 deg F**, and heat is lost on the exterior by convection to a fluid whose temperature is **40 deg F**. The convection film coefficient is **2E-4 BTU/sec-sq.in-F**. Calculate the stresses in the cylinder caused by the new temperature distribution.

This problem is **solved** in **two steps**. First, the **heat transfer** problem is modeled and solved, and the results of the heat transfer analysis (temperatures) are saved in a file, '**jobname.RTH**' (**R**esults **TH**ermal analysis).

Next the heat transfer boundary conditions and loads are removed from the mesh, the element type is changed from '**thermal**' to '**structural**', and the nodal temperatures that were saved in the file 'jobname.RTH' are recalled and applied as thermal loads.

1. Start ANSYS. Run Interactive > set (and remember the location of) your **WORKING DIRECTORY** and **jobname.** (Set the jobname to **Tutorial7C**. This step is **very important** since you will need to locate **Tutorial7C.RTH** in your working directory after the thermal analysis has been performed.)

2. Main Menu > Preferences > Select **Structural &Thermal > OK**

Enter the structural and thermal material properties, **E**, **prxy**, **alpx**, and **KXX**.

3. Main Menu > Preprocessor > Material Props > Material Models > Material Number 1 > Structural > Linear > Elastic > Isotropic > EX = 3.E7, PRXY = 0.3 > OK

Thermal Expansion > Secant Coefficient > Isotropic > ALPX = 6.5E-6 > OK

Thermal > Conductivity > Isotropic > KXX = 8.09 e-4 > OK > Close
4. Generate a quadrant of a section through the cylinder using ANSYS area creation tools. We will use **axisymmetric** modeling techniques.

At point $x = 5.0$, $y = 0.0$, create by corners a **5 by 10 inch rectangle**. Create a **partial annulus** at center $x = 0.0$, $y = 10.0$, R1 = 5, Theta1 = 0, R2 =10, Theta2 = 90. **Glue** the two areas together to form two four-edged regions suitable for mapped meshing. (**Modeling > Operate > Booleans > Glue > Areas**). Next select the thermal quad element.

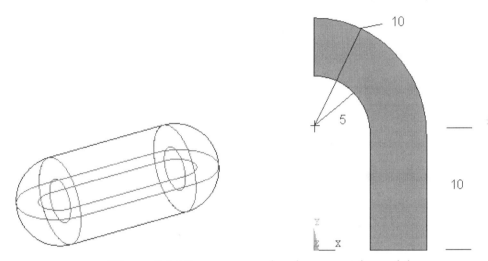

Figure 7-14 Pressure vessel and asymmetric model.

5. Main Menu > Preprocessor > Element Type > Add/Edit/Delete > Add > Thermal > Solid 8 node 77 > OK   (Scroll down to locate it.)

Options > Element behavior K3 Axisymmetric > OK > Close

6. Use mapped meshing to create the **5** by **20** mesh shown below.

7. Main Menu > Preprocessor > Loads > Define Loads > Apply > Thermal > Temperature > On Lines

Select the two lines on the interior and set the temperature to 75.

Figure 7-15 Cylinder mesh.

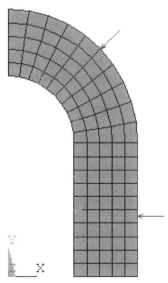

8. Main Menu > Preprocessor > Loads > Define Loads > Apply > Thermal > Convection > On Lines

Select the lines defining the outer surface and set the constant **Film coefficient = 2e-4** and the fluid **Bulk temperature = 40**.

9. Main Menu > Preprocessor > Loads > Define Loads > Apply > Thermal > Heat Flux > On Lines

Select the bottom horizontal line of symmetry and set the **heat flux** to zero.

Use **PlotCtrls > Surface Load Symbols > Heat Fluxes & Show pres as > Contours** to view the flux boundary conditions. Or **List > Loads > Surface > On All Lines**.

10. Main Menu > Solution > Solve > Current LS > OK

11. Main Menu > General Postprocessor > Read Results > First Set
Main Menu > General Postprocessor > Plot Results > Contour Plot > Nodal Solu > DOF Solution > Temperature > OK

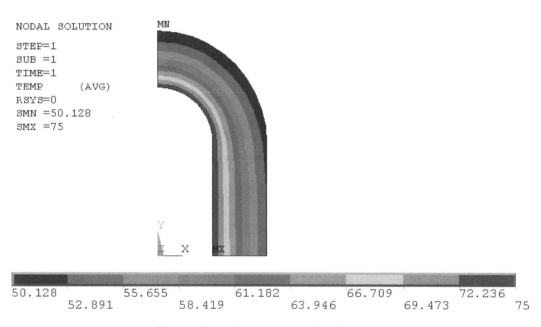

Figure 7-16 Temperature distribution.

The temperature on the interior is **75 deg F** as it should be and on the outside wall it is found to be about **50 F**. **Save the temperature results for use later.**

12. Utility Menu > File > Save Jobname.db (Check in your **working directory**.)

Delete all of the thermal loads (boundary conditions).

13. Main Menu > Preprocessor > Loads > Define Loads > Delete > All Load Data > All Loads & Opts. > OK

Switch the element type from thermal to structural.
14. Main Menu > Preprocessor > Element Type > Switch Elem Type (Select **Thermal** to **Struc**) **> OK** (This changes the element to the compatible 8 node structural element.)

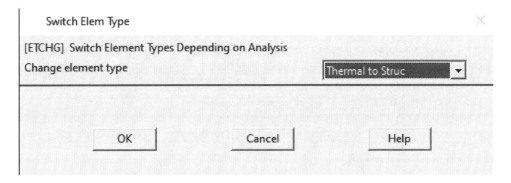

Figure 7-17 Switch element type to structural.

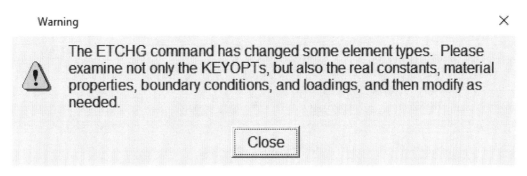

Figure 7-18 Warning of element change.

Close the warning message window and be sure to set the PLANE183 structural element options to **axisymmetric**. Another important **make** or **break** step.

15. Main Menu > Preprocessor > Element Type > Add/Edit/Delete > PLANE183 > Options > Element behavior K3 Axisymmetric > OK > Close. Plot > Elements

Now apply the structural boundary conditions and loads. (If the vessel had an internal and/or external pressure, it would be added here. For now we focus on the thermal effects only.)

16. Main Menu > Preprocessor > Loads > Define Loads > Apply > Structural > Displacement > On Lines

(Set UY = 0 on the bottom horizontal line of symmetry. You can use the Utility menu **List > Loads > DOF Constraints** or **PlotCtrls > Symbols > All Applied BCs** to check the model.)

Now read the nodal temperatures computed during the thermal analysis step.

17. Main Menu > Preprocessor > Loads > Define Loads > Apply > Temperature > From Thermal Analy. Browse and select **Tutorial7C.RTH.** (If you don't find it, check the current jobname by trying to change the jobname. Also use Utility Menu > List > Loads > Body > On All Nodes to check the thermal loads.)

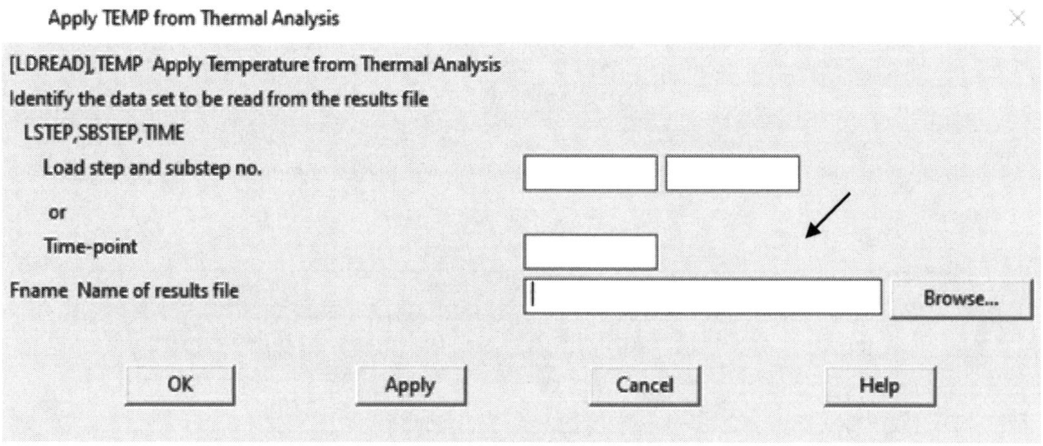

Figure 7-19 Enter name of thermal results file.

Solve the structural problem.

18. Main Menu > Solution > Solve > Current LS

19. Main Menu > General Postproc > Read Results > First Set > OK

20. Main Menu > General Postproc > Plot Results > Contour Plot > Element Solu > Stress > von Mises stress > OK

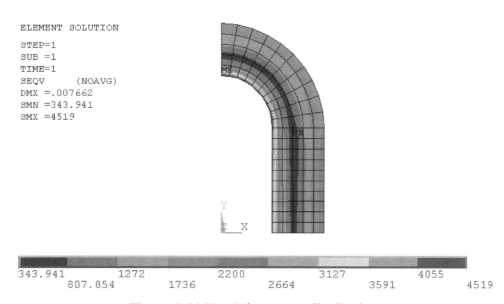

ELEMENT SOLUTION
STEP=1
SUB =1
TIME=1
SEQV   (NOAVG)
DMX =.007662
SMN =343.941
SMX =4519

| 343.941 | | 1272 | | 2200 | | 3127 | | 4055 | |
| 807.854 | | 1736 | | 2664 | | 3591 | | 4519 | |

Figure 7-20 Von Mises stress distribution.

The von Mises stress is seen to be a maximum of about 4500 psi in the end cap on the interior of the cylinder and would govern a yield-based design decision for a ductile steel.

Note that a finer mesh is generally required to get adequate accuracy for the stress analysis problem than is required for the thermal problem, so one must plan ahead in mesh design for the first phase of an analysis of the type discussed in this tutorial.

7-8 SUMMARY

The tutorials in this lesson illustrate the solution of thermal conduction/convection problems as well as the determination of thermal stresses arising from uniform or non-uniform temperature distributions. Although only planar and axisymmetric examples were considered, the extension to three-dimensional analysis follows directly from the examples here and those of Lesson 4.

7-9 PROBLEMS

7-1 Find the temperature distribution in the plate of Tutorial 2A if the hole is maintained at a temperature of 100 deg. C and the edges are exposed to a fluid at 0 deg C with a convection film coefficient of 20 W/m²-C. The material conductivity is 18 W/m-C. Assume no variation on the Z-direction so that a two-dimensional model is sufficient.

7-2 The shaft described in Problem 5-6 is subjected to a uniform 150 deg. F temperature rise. Use an axisymmetric model to find the maximum axial stress due to thermal effects only.

7-3 The pressure vessel of Lesson 3 is subjected to the thermal conditions of Tutorial 7C. Find the maximum von Mises stress due to thermal loading. Repeat with thermal and pressure loading.

7-4 Repeat 7-3 using the vessel of Problem 3-1.

7-5 Repeat Problem 7-1 but now assume the front and back surfaces of the plate are subject to convection the way fins on a cylindrical tube are used to cool (or heat) a fluid within the tube. Use three-dimensional modeling to determine the temperature distribution in the plate.

7-6 A composite wall is composed of two layers as shown. The conductivities are 0.05 W/cm-C for the 8 cm layer and 0.4 W/cm-C for the 3 cm layer. The free surface of the thick layer is maintained at 20 C. Convection at the free surface of the thin layer occurs with a film coefficient of 0.1 W/cm²-C and free stream temperature of –5 C.

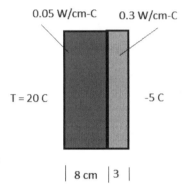

Find the temperature at the interface of the two layers. Use a two-dimensional model of rectangular elements of arbitrary y-direction height (say 1 cm) with zero heat flux in the y direction on the top and bottom boundaries.

Figure P7-6

7-7 An 8 in diameter steel disk 0.1 inches in thickness has a 2 in diameter hole located 1 inch radially from its center. The hole is maintained at 75 F and the exterior loses heat by convection to an environment whose temperature is 40 F. The convection film coefficient is 2.0E-4 BTU/sec-in²-F.

Find the temperature distribution and the value and location of the maximum von Mises stress, the maximum principal stress, and the maximum principal strain if the disk is initially in a stress-free state before the thermal conditions are applied.

Figure P7-7

7-8 Solve problem 7-2 using a three-dimensional model.

Lesson 8

Selected Topics

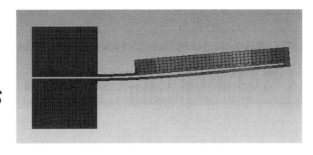

8-1 OVERVIEW

Selected topics not included in previous Lessons are included here for your reference.

8-2 LOADING DUE TO GRAVITY

The loading due to the weight of the structure in many of the examples we have analyzed so far was small in comparison to the applied loads and therefore was neglected in the analysis. If we want to consider the loading due to gravity, we need to make two additions to the input data: (1) Be sure to specify the **mass density** when entering material data and (2) Indicate the magnitude of the **acceleration of gravity** and its direction with respect to the global axis system for the problem.

To consider loading due to weight in Tutorial 1A for example, we would make the following additions to the text file. The length units for this analysis are inches, so we stick to inches. The **weight density** for steel is about 0.283 lbf/in^3. To find mass density, divide the weight density by the acceleration of gravity at the location on earth where the weighing took place, usually taken to be 386 in/sec^2 at sea level. Thus the **mass density** for steel in the British system inch units is

$$\rho = 0.283/386 = 0.0007331 \text{ lbf-sec}^2/\text{in}^4 = 0.7331\text{E-3 lbf-sec}^2/\text{in}^4. \quad \text{(about 7833 kg/m}^3\text{)}$$

A downward weight force on the shelf truss structure results from an upward acceleration equal to the acceleration of gravity. The *X*, *Y* and *Z* components of the acceleration of gravity are defined using the 'acel' command.

Add the following two data lines to the Tutorial 1A text file to include loading due to weight in the negative Y direction.

```
mp, dens, 1, 0.7331e-3          ! Material Property density
acel, 0, 386., 0                ! X, Y, Z components of gravity.
```

If the problem is being formulated by **interactive input**, the following selections and data inputs should be used.

When specifying the material properties, enter the mass density as well as the elastic modulus and Poisson's ratio. Double click on Density and enter its value.

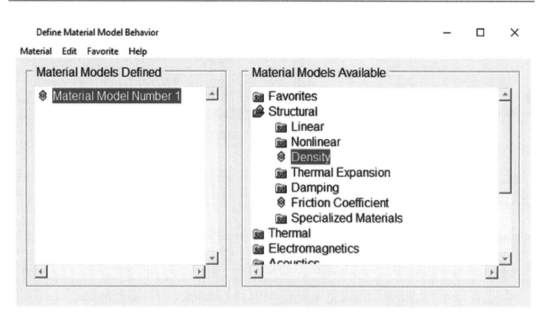

Figure 8-1 Mass density input.

To enter the acceleration of gravity,

Main Menu > Preprocessor > Loads > Define Loads > Apply > Structural > Inertia > Gravity > Global Enter Y-comp **386.** > **OK**.

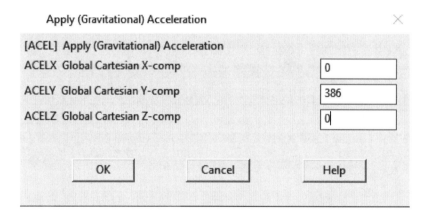

Figure 8-2 Acceleration of gravity input.

8-3 MULTIPLE LOAD CASES

Often we want to know the response of a system to more than one load condition. In the shelf truss problem for example, we might need to analyze the behavior under a side loading condition together with the downward load. Or the downward load might be considered by itself as one case, and the combination of downward and side loads as a second case. The support movement condition could then be added as a third case. There is no need to reformulate the problem three times to consider these combinations. The basic structural model does not change in any way, just the loadings.

ANSYS allows multiple load cases to be considered through the use of multiple load steps, a different step for each case. During preprocessing, each load case is defined and then written to a file to be used during the solution. In the solution phase this file is read and all load cases treated, not just the 'Current Load Step'. And finally in postprocessing, each load case (step) is viewed independently.

For Tutorial 1A with **fx = 0., fy = –300 lbf** at node 2 as **case one** and **fx = 100, fy = 0** at node 2 as **case two**, the input text file would be:

```
f, 2, fx, 0.
f, 2, fy, -300.
lswrite                 ! write load step to file

f, 2, fx, 100.
f, 2, fy, 0.
lswrite

/solu
antype,static

lssolve,1,2             ! Initiates solution for load step files 1 and 2
finish
```

The **interactive** commands for considering **multiple load cases** are: In the Preprocessor, use **Loads > Define Loads > Apply** to define the first load case as described earlier then select

Main Menu > Preprocessor > Loads > Load Step Opts > Write LS File (Give the load case a number) **> OK**

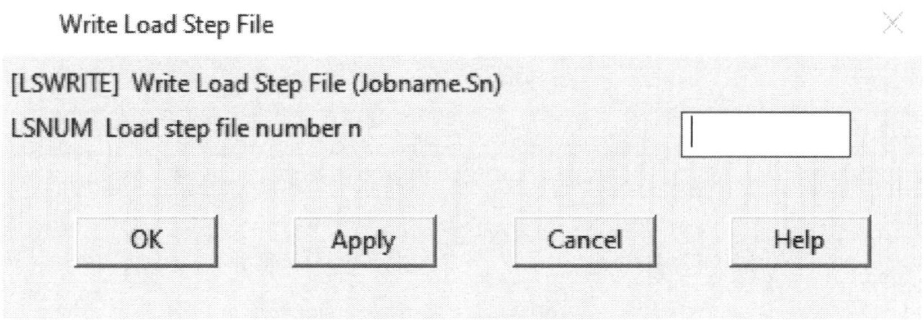

Figure 8-3 Write Load Step File.

Define the next load set (be sure to delete any loading not needed for the second set) and repeat the above.

Then read these load set files during the solution step.

Main Menu > Solution > Solve > From LS Files . . . (enter starting, ending, and increment) **> OK**

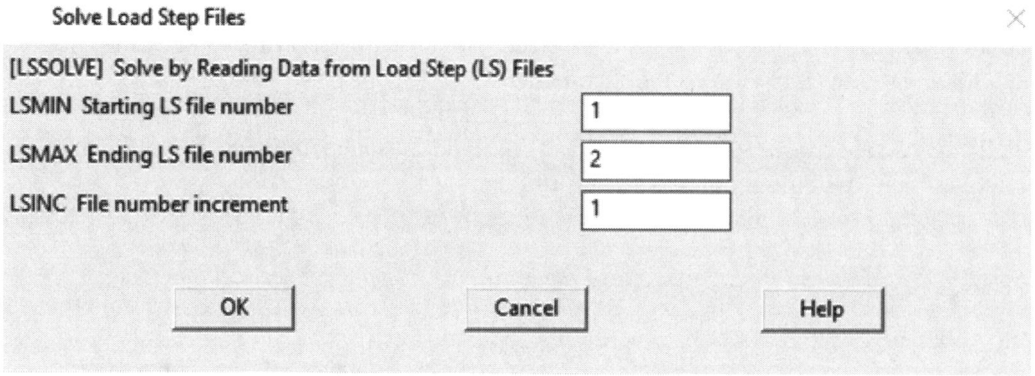

Solve Load Step Files   ✕

[LSSOLVE] Solve by Reading Data from Load Step (LS) Files

LSMIN Starting LS file number | 1

LSMAX Ending LS file number | 2

LSINC File number increment | 1

| OK | | Cancel | | Help |

Figure 8-4 Solve load step files.

To evaluate the computed results,

Main Menu > GeneralPostproc > Read Results > First Set > Plot Results etc.

Then

Main Menu > GeneralPostproc > Read Results > Next Set > Plot Results etc.

Continue reading calculated results data sets, plotting, listing, etc until you have all of the results desired from the analysis.

8-4 ANSYS MATERIALS DATABASE

We have entered the materials data for all of the examples thus far, but ANSYS also provides a database of commonly used materials. (Access details depend on your local installation.)

Main Menu > Preprocessor > Material Props > Material Library > Select Units from **SI, CGS, BFT, BIN,** or **USER > OK.**

Main Menu > Preprocessor > Material Props > Material Library > Import Library

Enter file names or browse (On Windows systems Program Files\Ansys Inc\ANSYS Student\V231\ANSYS\matlib) to find a data file in the ANSYS Matlib directory. Specify the material number **MAT** (1, 2, 3, etc.) for your model > **OK.**

The data file is displayed on the screen and is incorporated in your model database. To verify **List > Properties > All Materials.**

Data for **304 steel** in British system Inch units (BIN), for example, is shown below.

```
! ANSYS $RCSfile: Stl_AISI-304.BIN_MPL,v $
! Modified on $Date: 2009/11/13 16:18:22 $
! Source ID = $Revision: 1.3 $
/COM,Typical material properties for DEMO purposes only
/NOP
/COM,Internal UNITS set at file creation time = BIN
TBDEL,ALL,_MATL
MPDEL,ALL,_MATL
MPTEMP,R5.0, 1, 1, 0.000000000E+00,
MPDATA,R5.0, 1,EX  ,_MATL , 1,  27992720.0      ,
MPTEMP,R5.0, 1, 1, 0.000000000E+00,
MPDATA,R5.0, 1,NUXY,_MATL , 1, 0.290000000     ,
MPTEMP,R5.0, 1, 1, 0.000000000E+00,
MPDATA,R5.0, 1,ALPX,_MATL , 1, 9.888888889E-06,
MPTEMP,R5.0, 1, 1, 0.000000000E+00,
MPDATA,R5.0, 1,DENS,_MATL , 1, 7.514795200E-04,
MPTEMP,R5.0, 1, 1, 0.000000000E+00,
MPDATA,R5.0, 1,KXX ,_MATL , 1, 2.182244000E-04,
MPTEMP,R5.0, 1, 1, 0.000000000E+00,
MPDATA,R5.0, 1,C   ,_MATL , 1,  46.2864080     ,
/GO
```

8-5 ANSYS HELP

A wealth of on-line help information is provided with the ANSYS software. Select **Help** > **Help Topics** from the Utility menu and the following information screen is displayed.

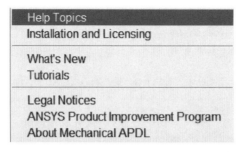

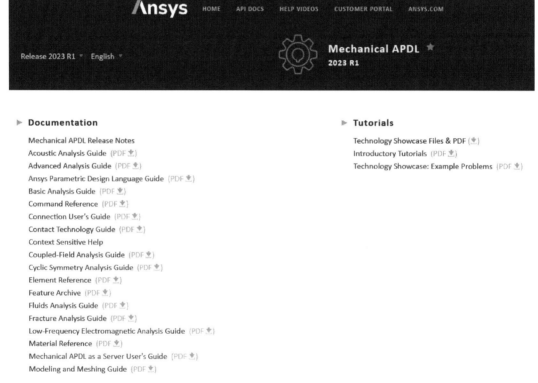

Figure 8-5 ANSYS help facility.

The contents shown include the structural analysis guide and the thermal analysis guide as well as an extensive set of very helpful **verification problems** and their data files. These are labeled VMXX, where XX is the verification problem number.

Select **Static Structural Analysis > 2.5 Where to Find Other Examples**

A search of the references can also be performed using keyword search methods.

8-6 CONSTRAINTS

Often one wants to add constraints to the model in order to enforce certain conditions not inherent in the modeling discussed previously. One of these is the introduction of a **release** or **hinge** where two elements join, particularly useful for beam modeling.

Another involves modeling of **rigid** objects, which are not always properly represented by using a very, very stiff element because that can sometimes produce undesirable numerical effects.

Look for information on '**coupled sets**' in the documentation for more help with constraints.

8-7 NATURAL FREQUENCIES

Natural frequencies of systems are those frequencies at which resonant response occurs under the right excitation conditions. Knowledge of these critical dynamic frequencies is an essential step in the design or evaluation of a system subjected to dynamic loadings.

8-8   TUTORIAL 8A – CANTILEVERED BEAM FREQUENCIES

In this first tutorial we compute the natural frequencies of vibration of a long slender bar that has pinned supports at both ends.

Consider a **steel** bar **0.5 inch in thickness, 1.0 inch high**, and **10 inches in length.** See the figure below. We will calculate the first **four bending frequencies** of this beam.

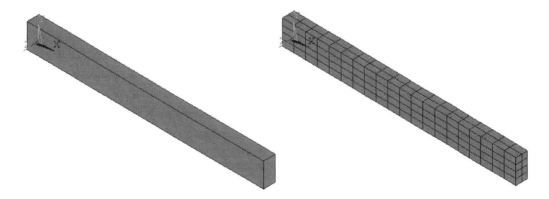

Figure 8-6 Cantilever beam.

The **global stiffness** and **global mass** matrices for a structure can be used to find the natural frequencies of vibration of the structure. The mass matrix will be available if the **mass density** is supplied along with the other material properties (be sure you get the units right). ANSYS has a number of solution methods available to compute natural frequencies and modes of vibration. The solver that is selected depends upon the size of the problem under consideration and the number of frequencies desired as well as other factors. The calculated results can **depend very strongly** on the number and spacing of the nodes and elements used in the model.

Natural frequency calculation in ANSYS is called **MODAL ANALYSIS**; to extract the natural frequencies using the default method, use the steps outlined below.

1. Create the **0.5 x 1.0 x 10 volume** using a **solid modeler** or **ANSYS geometry commands**.

2. Select the **solid element Brick 20node 186**.

3. Use **mapped meshing commands** to create a **2 x 4 x 20 element mesh**.

4. Specify the **Elastic Modulus, Poisson's Ratio** and **Mass Density** for Steel or use the materials database. (IPS Mass density (0.283 lbf/in^3)/386 in/sec^2 = 0.000733)

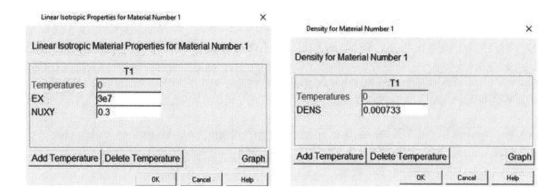

Figure 8-7 Material properties.

5. Fix all degrees of freedom on the **left end** of the beam.

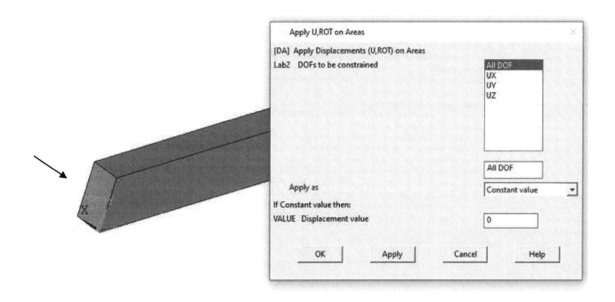

Figure 8-8 Boundary conditions.

6. Main Menu > Solution > Analysis Type > New Analysis > Modal > OK

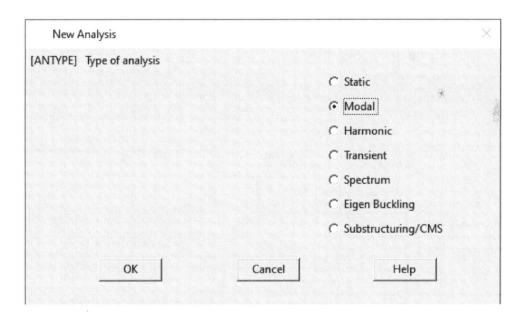

Figure 8-9 Select Modal Analysis.

7. Analysis Options > Block Lanczos, No. of Modes to Extract (Enter 4.), **Expand mode shapes > yes, NMODE No. modes to expand** (Enter 4), **OK, FREQB** 0, **FREQE** 0   Accept the defaults, Normalize modes **To Mass Matrix** (Gives orthonormal modes.) **> OK**

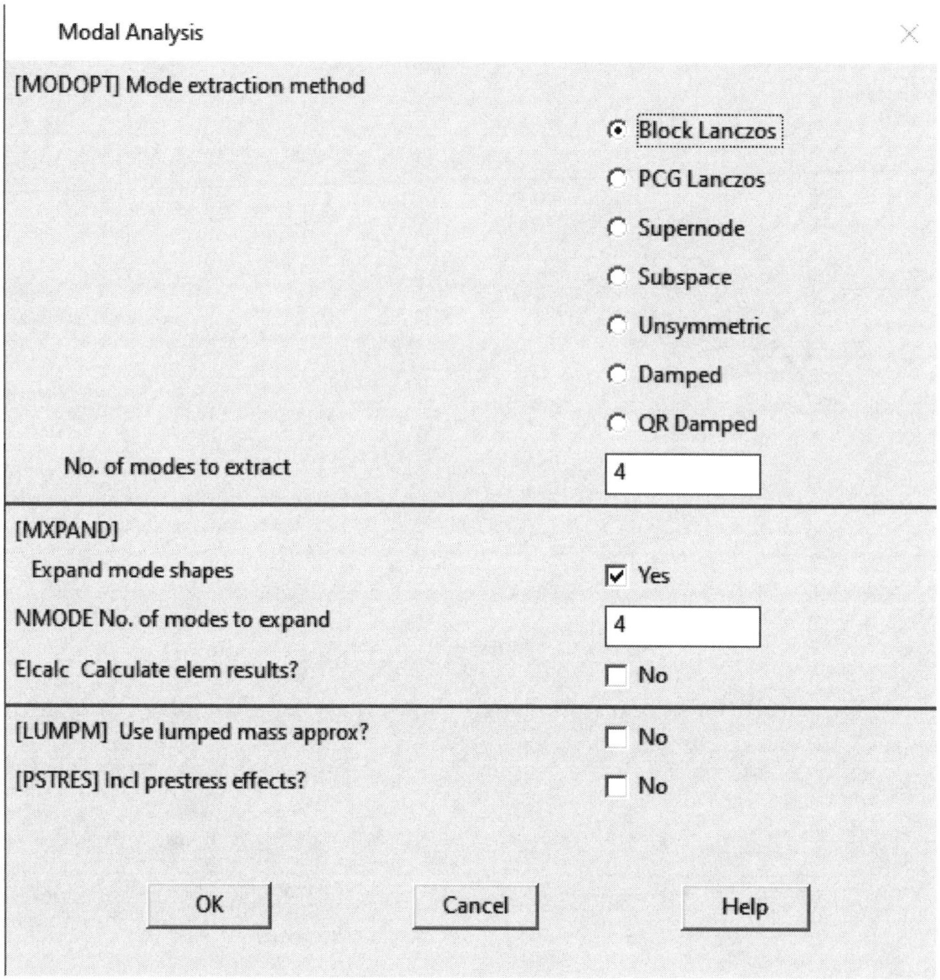

Modal Analysis   ✕

[MODOPT] Mode extraction method

                                                ⦿ Block Lanczos

                                               ◯ PCG Lanczos

                                               ◯ Supernode

                                               ◯ Subspace

                                               ◯ Unsymmetric

                                               ◯ Damped

                                               ◯ QR Damped

        No. of modes to extract   `4`

[MXPAND]

  Expand mode shapes   ☑ Yes

NMODE No. of modes to expand   `4`

Elcalc Calculate elem results?   ☐ No

[LUMPM] Use lumped mass approx?   ☐ No

[PSTRES] Incl prestress effects?   ☐ No

          OK   Cancel   Help

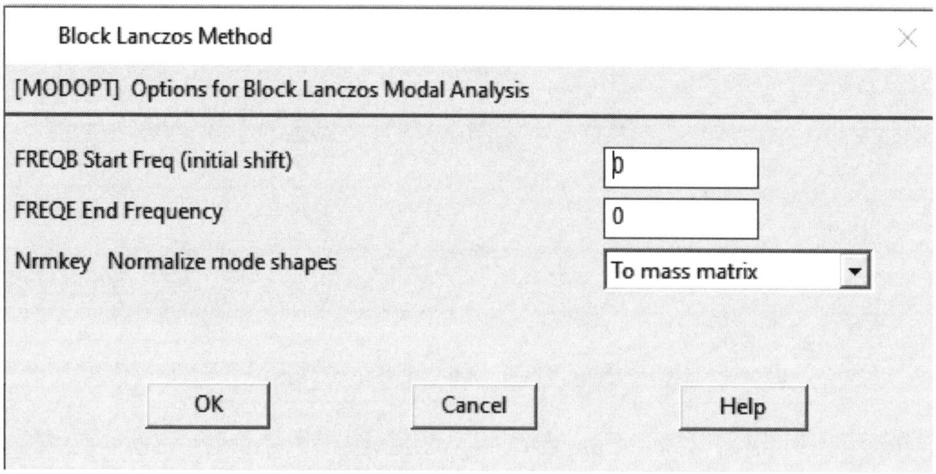

Block Lanczos Method   ✕

[MODOPT] Options for Block Lanczos Modal Analysis

FREQB Start Freq (initial shift)   `0`

FREQE End Frequency   `0`

Nrmkey Normalize mode shapes   `To mass matrix ▼`

          OK   Cancel   Help

Figure 8-10 Analysis Options.

Calculate the natural frequencies.

8. Main Menu > Solution > Solve > Current LS > OK

Evaluate the results.

9. Main Menu > General Postprocessor > Results Summary (Displays the calculated frequencies in Hz.)

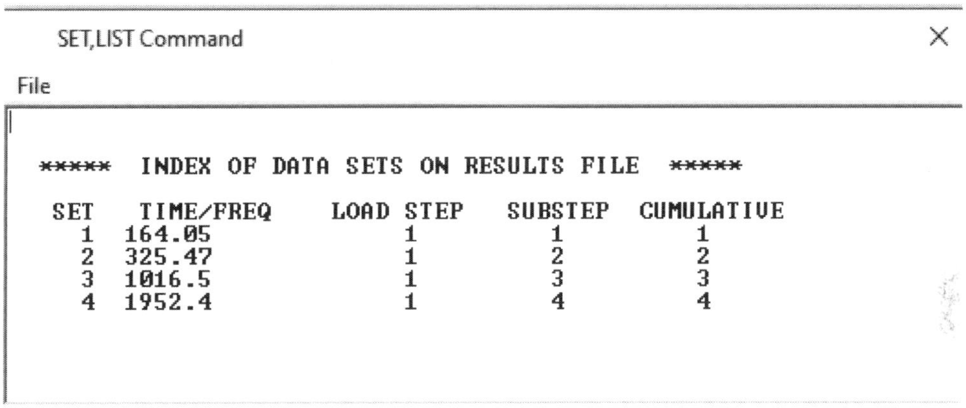

```
 SET,LIST Command                                                    ×

File

 *****    INDEX OF DATA SETS ON RESULTS FILE    *****

  SET     TIME/FREQ     LOAD STEP     SUBSTEP     CUMULATIVE
   1     164.05             1            1             1
   2     325.47             1            2             2
   3     1016.5             1            3             3
   4     1952.4             1            4             4
```

Figure 8-11 First four calculated natural frequencies.

Observe the computed mode shapes.

10. Main Menu > General Postprocessor > Read Results > First Set

11. Main Menu > General Postprocessor > Plot Results > Deformed Shape > Def. + Undef. The **first mode** of vibration is shown.

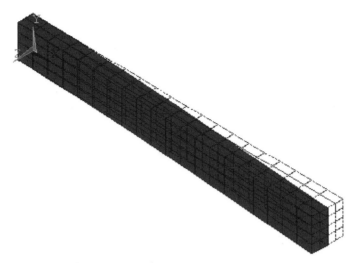

Figure 8-12 First mode of vibration.

It is also very helpful to use **PlotCtrls > Animate** to visualize the motion. The 'raise hidden' button near the top of the screen can be used to display the **Animation Controller** window to manipulate the display.

12. Use **List Results > Nodal Solution > DOF solution > All DOFs** to list the calculated mode shape values.

For the next mode **Read Results > Next Set**

Plot Results > Deformed Shape > Def. + Undef., Continue for all modes of interest.

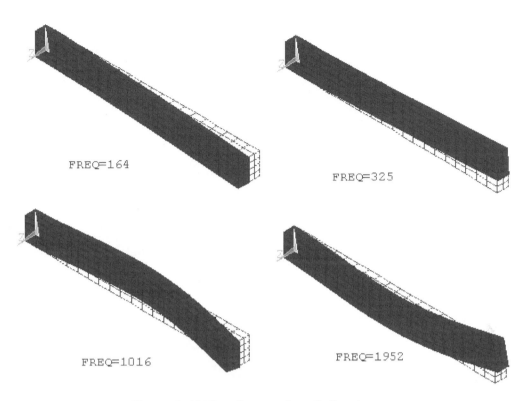

Figure 8-13 First four modes of vibration.

Note that the computed frequencies are given in cycles/second (Hz). Modes can be normalized with the largest displacement value set to 1.0 or normalized on the mass matrix so that the generalized mass terms are all 1.0 (orthonormal modes).

Next we clear the mesh and create a finer mesh to access the effect of mesh density on the results. Results from analysis of a 4 x 8 x 40 mesh of 1280 elements are shown below.

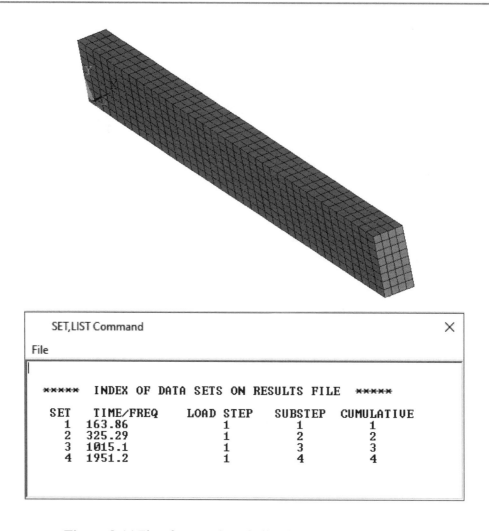

Figure 8-14 First four modes of vibration using second mesh.

Note that the results above are only slightly different from those computed earlier. (Also note that these results may not compare too well with elementary beam vibration theory which analyzes an object with a little higher aspect ratio, length/width, than the 10 to 1 in this case.)

8-9 ANSYS Files

A number of permanent and temporary files are used by ANSYS during the solution of a typical problem. These have the form *.ext where * is the Jobname and ext is an extension which is used to identify the data stored in the file. If ANSYS starts with a blank Jobname, the default name is set to '**file**'.

Some files are text (ASCII) files and some are binary files. Among the most important are the following:

*.db - A binary file that stores the job database. It can contain model data (geometry, materials, elements, etc.), model data plus solution results, or model, solution and postprocessing data. The exact contents depend upon how the db file is saved. The **File > Save as Jobname.db** (or **Save as**) command will save the model and solution results if the solution has been performed. The **QUIT (Exit) > Save . . .** provides options to save part, all or none of the data currently in memory.

*.rst - The binary file that contains the results of a structural analysis and is created when the solution step is performed.

*.rth - The binary results file for a thermal analysis.

*.log - A text file that contains all the command instructions from a session with ANSYS. It can be used to recreate a session by using the **Read Input from** menu choice to read commands contained in the log file. New sessions append to the existing log file when ANSYS starts, so using *.log as input requires that you cut and paste to select the session in question or plan ahead and delete *.log before starting a session whose commands you want to preserve for later use. The log file provides a way to convert an interactive session into a text file definition of a specific analysis. CDWRITE is another way of doing this but uses the db file.

*.err - A text file containing error and warning messages written during an analysis.

*.dbb - Binary file which is a copy of the db file.

To retrieve a model, all that is needed is the db file. The db file will contain model, model and solution, or model, solution, and postprocessing data depending upon how it was saved.

8-10 SUMMARY

The number of possible ANSYS modeling and solution options is extremely large, and this lesson has presented a few that may be of interest.

NOTES:

INDEX

REFERENCES

Books

Esam M. Alawadhi, *Finite Element Simulations Using ANSYS 2nd Edition*, T&F INDIA, 2019.

Arshad Ali and Zeeshan Azmat, *Analysis of Composite Structure under Thermal Load Using ANSYS*, LAP Lambert Academic Publishing, 2010, 84 pp.

R.B. Choudary, *ANSYS 2020: Structural Analysis Using the ANSYS Mechanical APDL Release 2020 R1 Environment*, Dreamtech Press, 2022.

Michael R. Hatch, *Vibration Simulation Using MATLAB and ANSYS*, Chapman and Hall/CRC, 2000, 654 pp.

Erdogan Madenci, Ibrahim Guven, Bahattin Kilic, *Fatigue Life Prediction of Solder Joints in Electronic Packages with ANSYS*, The Springer International Series in Engineering and Computer Science, Springer, 2002, 208 pp.

Erdogan Madenci and Ibrahim Guven, *The Finite Element Method and Applications in Engineering Using ANSYS*, Springer, 2015, 671 pp.

Saeed Moaveni, *Finite Element Analysis Theory and Application with ANSYS*, 4th Ed, Prentice Hall, 2014, 936 pp.

Tadeusz Stolarski and Y. Nakasone, *Engineering Analysis with ANSYS Software*, 2nd edition, Butterworth-Heinemann, 2018, 562 pp.

Sham Tickoo, *ANSYS 11.0 for Designers*, CADCIM Technologies, 2009, 544 pp.

Web sites: See many from a web search using 'ANSYS'

NOTES: